PRACTICAL SAFETY MANAGEMENT SYSTEMS

A practical guide to transform your safety program
into a functioning safety management system

PAUL R. SNYDER & GARY M. ULLRICH

Paul Alexiou

AVIATION SUPPLIES & ACADEMICS, INC.
NEWCASTLE, WASHINGTON

Practical Safety Management Systems
by Paul R. Snyder and Gary M. Ullrich

Aviation Supplies & Academics, Inc.
7005 132nd Place SE
Newcastle, Washington 98059-3153
asa@asa2fly.com | www.asa2fly.com

Published 2016 by Aviation Supplies & Academics, Inc.

Front cover photo: ©Image Source (Fotolia)
Back cover photos (from top): ©Chalabala (iStock); ©huseyintuncer (iStock); ©Chalabala (iStock); Courtesy Freefly Systems Inc. (www.freeflysystems.com); ©JuNiArt (iStock); ©Chalabala (iStock)

ASA-SMS
ISBN 978-1-61954-424-6

Printed in the United States of America

2020   2019   2018   2017   2016   9 8 7 6 5 4 3 2 1

Additional information and resources to support your studies can be found on the "Reader Resources" page for this book by visiting **www.asa2fly.com/reader/SMS**

CONTENTS

INTRODUCTION v

CHAPTER 1 1
Overview and History of Safety Management Systems

CHAPTER 2 13
Understanding the Components of an SMS

CHAPTER 3 39
SMS Costs Versus Benefits

CHAPTER 4 49
Safety Management Systems Versus Safety Programs

CHAPTER 5 57
Scalability of SMS

CHAPTER 6 69
Basic Safety Concepts

CHAPTER 7 91
SMS Planning and Process

CHAPTER 8 111
Transitioning Your Safety Program to a Safety Management System

CHAPTER 9 117
Developing a Safety Policy for Your Organization

CHAPTER 10 129
Safety Risk Management

CHAPTER 11 149
Safety Assurance and Continuous Monitoring

CHAPTER 12 159
Safety Assurance and Audits

CHAPTER 13 169
SMS and Your Safety Culture

CHAPTER 14 181
Creating Your SMS Manual

GLOSSARY 193

INDEX 203

INTRODUCTION

PRACTICAL SAFETY MANAGEMENT SYSTEMS provides an up-to-date practical guide to transform your safety program, regardless of size and scope, into a functioning safety management system (SMS). This book moves beyond the theoretical discussion and engages you through hands-on exercises designed to apply SMS concepts and principles.

The International Civil Aviation Organization (ICAO) requires that an SMS address safety risk in air operations, maintenance, air traffic services and airports. These ICAO requirements have been expanded to include flight training and design and production of aircraft. SMS has literally affected all areas of the aviation industry worldwide, including independent contractors who provide services to the aviation industry.

The FAA published the long awaited 14 CFR Part 5 "Safety Management Systems" on January 8, 2015. After the standard 60-day waiting period, the rule became effective March 9, 2015, requiring all Part 121 certificate holders to develop and implement an SMS within their organizations. Additionally, the new Part 5 cancelled the SMS Pilot Project and created a new SMS Voluntary Program (SMSVP) using the SMSVP Standard. The SMSVP meets "state" recognition requirements as defined in ICAO Annex 6. Any organization that wants to voluntarily implement an SMS Program must use the SMSVP Standard if they want to implement an SMS accepted by the FAA. This textbook covers the requirements for both Part 5 and the SMSVP Standard.

The newly created SMS Voluntary Program (SMSVP) Guide, AFS-900-002-G201, deleted the five SMS Pilot Program Levels and created 3 new SMSVP Levels:

1. **SMSVP Active Applicant.** The certificate holder and Certificate Management Team have committed to sufficiently support the SMS implementation and validation processes.
2. **SMSVP Active Participant.** The certificate holder officially begins and maintains its implementation efforts. The certificate holder receives an acknowledgment letter.
3. **SMSVP Active Conformance.** The Certificate Management Team (CMT) and the SMS Program Office (SMSPO), located in Washington D.C., acknowledge full implementation of the certificate holder's SMS. The certificate holder receives an acknowledgment letter. The certificate holder is expected to use and continually improve its safety management processes.

Many profess to have an understanding of SMS, yet very few have had practical hands-on experience to apply SMS concepts for their particular organization. SMS must be tailored to meet the size and scope of each organization. SMS is a systematic approach to managing safety, including the necessary organizational structure, accountabilities, policies, and processes of each organization. It cannot be purchased from an outside vendor and placed on a shelf, but must be adapted to each organization and continuously improved to meet the mission of the organization while reducing risk to the lowest practical level.

ABOUT THE AUTHORS

Paul R. Snyder

Professor Snyder is an Assistant Professor for the John D. Odegard School of Aerospace Sciences at the University of North Dakota. He continues to be active in flight operations as an FAA Designated Pilot and Instructor Examiner, Insitu-Certified ScanEagle UAS pilot, and a UND Chief Flight Instructor. As a DOT-certified SMS trainer, Professor Snyder is an active member of UND Executive SMS committee and Events Review Team (ERT), analyzing flight data and conducting safety risk assessments to reduce risk within the organization. He has been a leader in working with the FAA to be recognized as meeting Active Conformance for UND's Part 141 Pilot School and Part 145 Maintenance Repair Station under the FAA Pilot Study Program.

Professor Snyder also actively teaches Safety Management System (SMS) and advanced flight courses and training seminars for the University of North Dakota. Past research activities include FAA Industry Training Standards (FITS)—Scenario Based Training, UAS curriculum design, UAS detect and avoid, UAS pilot training, and helicopter approach and landing lighting systems. He holds a degree in Aeronautical Studies and Master of Science in Educational Leadership.

Gary M. Ullrich

Before joining the University of North Dakota, Professor Ullrich was a Test Group Pilot for FlightSafety, Adjunct Assistant Professor for Embry Riddle Aeronautical University, and Chief of Safety/Instructor/Evaluator Pilot with the United States Air Force. While working at FlightSafety, he helped to create their ISO 9001 Program.

Professor Ullrich joined UND Aerospace in 2006. In 2014, he helped UND Aerospace achieve their FAA-recognized SMS Level 3 status. He currently teaches Aviation Safety, Safety Management Systems, Aircraft Accident Investigation, Advanced Aerodynamics, and Long Range Navigation and International Procedures. He is a commercial pilot with multi-engine, instrument, and Boeing 720/707 type ratings.

CHAPTER 1
Overview and History of Safety Management Systems

OBJECTIVES

- Define the definition of Safety Management System (SMS).
- Recall the history which led to the international requirement for an SMS Program.
- Describe the United States statuary requirement to establish an SMS Program.
- Explain the history of the SMS Pilot Programs.
- Summarize the important parts of 14 CFR Part 5.
- List and define the three levels of the SMS Voluntary Program (SMSVP).
- Recall the four components of SMS.

KEY TERMS

- 14 CFR Part 5
- ICAO Annex
- ICAO Standards and Recommended Practices (SARPs)
- International Civil Aviation Organization (ICAO)
- Safety Assurance (SA)
- Safety Management System (SMS)
- Safety Policy
- Safety Promotion
- Safety Risk Management (SRM)
- SMS Pilot Project
- SMS Voluntary Program (SMSVP)
- SMSVP Active Applicant
- SMSVP Active Conformance
- SMSVP Active Participant

WHAT IS SMS?

Safety management system (SMS) is the formal, top-down business-like approach to managing safety risk, which includes a systemic approach to managing safety, including the necessary organizational structures, accountabilities, policies and procedures.

SMS is becoming a standard throughout the aviation industry worldwide. It is recognized by the **International Civil Aviation Organization (ICAO)**, civil aviation authorities (CAA), the Interagency Planning Office, and product/service providers as the next step in the evolution of safety in aviation. SMS is also becoming a standard for the management of safety beyond aviation. Similar management systems are used in the management of other critical areas such as quality, occupational safety and health, security, and environment.

Safety management systems for product/service providers (certificate holders) and regulators integrates modern safety risk management and safety assurance concepts into repeatable, proactive systems. SMSs emphasize safety management as a fundamental business process to be considered in the same manner as other aspects of business management.

By recognizing the organization's role in accident prevention, an SMS provides to both certificate holders and FAA:

- A structured means of safety risk management **decision making**.
- A means of demonstrating safety **management capability** before system failures occur.
- Increased confidence in **risk controls** though structured **safety assurance** processes.
- An effective interface for **knowledge sharing** between regulator and certificate holder.
- A **safety promotion** framework to support a sound **safety culture**.

Technology and system improvements have made great contributions to safety. That said, part of being safe is about attitudes and paying attention to what your surroundings are telling you. Whether through data or through the input of employees and others, recognizing that many opportunities exist to stop an accident is the first step in moving from reactive to predictive thinking.

Safety begins from both the top down and the bottom up. Everyone from the receptionist, ramp worker, pilot, supervisors, managers, the chief executive officer (CEO), and FAA Inspector has a role to perform.

SMS is all about safety-related decision-making throughout the entire organization. Thus it is a decision-maker's tool, not a traditional safety program separate and distinct from business and operational decision-making.

SMSs can be a complex topic with many aspects to consider, but the defining characteristic of an SMS is that it is a decision-making system. An SMS does not have to be an extensive, expensive, or sophisticated array of techniques in order to do what it is supposed to do. Rather, an SMS is built by structuring safety management around four components:

- **Safety policy;**
- **Safety risk management (SRM);**
- **Safety assurance (SA);** and
- **Safety promotion.**

Safety Policy

Safety policy consists of setting objectives, assigning responsibilities, and setting standards. It is also where management conveys its commitment to the safety performance of the organization to its employees. As SRM and SA processes are developed, the organization should come back to the safety policy to ensure that the commitments in the policy are being realized and that the standards are being upheld.

Safety Risk Management

The SRM component provides a decision-making process for identifying hazards and mitigating risk based on a thorough understanding of the organization's systems and their operating environment. SRM includes decision making regarding management acceptance of risk to operations. The SRM component is the organization's way of fulfilling its commitment to consider risk in their operations and to reduce it to as low a level as possible. In that sense, SRM is a design process, a way to incorporate risk controls into processes, products, and services or to redesign controls where existing ones are not meeting the organization's needs.

Safety Assurance

SA provides the organization with the necessary processes to give confidence that the systems meet the organization's safety objectives and that mitigations or risk controls developed under SRM are working. In SA, the goal is to watch what is going on and review what has happened to ensure that objectives are being met. Thus, SA requires monitoring and measuring safety performance of operational processes and continuously improving the level of safety performance. Strong SA and safety data analysis processes yield information used to maintain the integrity of risk controls. SA processes are thus a means of ensuring the safety performance of the organization, keeping it on track, correcting it where necessary, and identifying needs for rethinking existing processes.

Safety Promotion

The last component, safety promotion, is designed to ensure that an organization's employees have a solid foundation regarding their safety responsibilities, the organization's safety policies and expectations, reporting procedures, and a familiarity with risk controls. Training and communication are the two key areas of safety promotion.

An SMS does not have to be large, complex, or expensive in order to add value. If there is active involvement by the operational leaders, open lines of communication up and down the organization and among peers, vigilance in looking for new operations, and assurance that employees know that safety is an essential part of their job performance, the organization will have an effective SMS that helps decision makers at all levels.

EVOLUTION OF SAFETY MANAGEMENT

Safety management systems are the product of a continuing evolution in aviation safety. Early aviation pioneers had little safety regulation, practical experience, or engineering knowledge to guide them. Over time, careful regulation of aviation activities, operational experience, and improvements in technology have contributed to significant gains in safety. In the next major phase of improvement to safety, a focus on individual and crew performance or "human factors" further reduced accidents.

The history of the progress in aviation safety can be divided into three areas: the technical era, human factors era, and the organizational era (Figure 1-1).

Figure 1-1 Evolution of safety.

The Technical Era—From the Early 1900s Until the Late 1960s

Aviation emerged as a form of mass transportation where identified safety deficiencies were initially related to technical factors and technological failures. The focus of safety endeavors was placed on the investigation and improvement of technical factors. By the 1950s technological improvements led to a gradual decline in the frequency of accidents and safety processes were broadened to encompass regulatory compliance and oversight.

The Human Factors Era—From the Early 1970s Until the Mid-1990s

In the early 1970s, the frequency of aviation accidents was significantly reduced because of major technological advances and enhanced safety regulations. Aviation became a safer mode of transportation, and the focus of safety endeavors was extended to include human factors, particularly the man/machine interface.

Despite the investment of resources in error mitigation, human performance continued to be cited as a recurring factor in accidents. The application of human factors science tended to focus on the individual, without fully considering the operational and organizational context. It was not until the early 1990s that it was first acknowledged that individuals operate in a complex environment, which includes multiple factors having the potential to affect behavior.

The Organizational Era—From the Mid-1990s to the Present Day

During the organizational era safety began to be viewed from a systemic perspective, encompassing organizational factors in addition to human and technical factors. The notion of the organizational accident was introduced. It considered the impact of organizational culture and policies on the effectiveness of safety risk controls. Additionally, a new, proactive approach to safety supplemented traditional data collection and analysis efforts, which had been limited to the use of data collected through investigation of accidents and serious incidents. This new approach is based on routine collection and analysis of data using both proactive as well as reactive methodologies, meant to monitor known safety risks and detect emerging safety issues. This monitor and detect logic is the core of a modern SMS approach.

Each approach has led to significant gains in safety. Even with these significant advances, however, we still have opportunities to take preventative action against accidents. The question for the aviation community is, "what is the next step?"

Careful analysis typically reveals multiple opportunities for actions that could have broken the chain of events and possibly prevented an accident. These opportunities represent the organization's role in accident prevention. The term "organizational accident" was developed to describe accidents that have causal factors related to organizational decisions and attitudes. SMS is an approach to improving safety at the organizational level.

WHY DO WE NEED SMS?

We are now in a position where the "common cause" accidents are diminishing in number. While it's a major success story, it's not a place to rest. When we find a cause that affects all or part of a large population of operators or other aviation participants, we can address risk through rulemaking—a risk control that applies to everyone to address risks to which everyone is exposed. There will always be some of these risks and work will continue to find them and address them.

Many accidents that occur, however, are caused by the unique aspects of the operating environments of individual operators in narrow segments of the aviation community. The causal factors of these accidents aren't common to everyone; they must be found and addressed with methods that are sensitive to the nuances of the individual operator's situation. One of the defining characteristics of an SMS is its emphasis on risk management within the individual operator's environment and situation—it's a gap filler between the common cause risk factors that are addressed by traditional regulations and those that are more elusive.

HOW SMS ADDRESSES AN ORGANIZATION'S ROLE IN SAFETY

SMS requires an organization to examine its operations and the decisions around those operations. SMS allows an organization to adapt to change, increasing complexity, and limited resources. SMS also promotes the continuous improvement of safety through specific methods to predict hazards from employee reports and data collection. Organizations then use this information to analyze, assess, and control risk. Part of the process includes the monitoring of controls and of the system itself for effectiveness. SMS helps organizations comply with existing regulations while predicting the need for future action by sharing knowledge and information. Finally, SMS includes requirements that enhance the safety attitudes of an organization by changing the safety culture of leadership, management, and employees. All these changes are designed to move an organization from reactive thinking to predictive thinking.

FLIGHT STANDARDS VOLUNTARY SMS PILOT PROJECTS (2007-2015)

The FAA Flight Standards organization conducted voluntary **SMS Pilot Projects** for external SMSs, specifically for operators and product/service providers. Approximately 130 voluntary pilot project participants realized substantial safety and financial benefits through their development of a voluntary SMS.

There were three primary objectives of the SMS Pilot Projects:

- **Develop implementation strategies**;
- **Develop oversight interfaces**; and
- **Gain experience for FAA and service providers**.

Members of the Voluntary SMS Pilot Project provided a two-way communication mechanism between the SMS Program Office and participants in voluntary implementation. It also provided a forum for knowledge sharing among participants. Pilot Project participants shared best practices and Lessons Learned with each other and the FAA SMS Program Office.

The SMS Pilot Project contained five levels (Figure 1-2) for the SMS journey. Level 0 was the entry point. Level 0 was not so much a level as a status. It indicated that the service provider has not started formal SMS development or implementation. It indicated the time period between a service provider's first request for information from the FAA on SMS implementation and when the service provider's top management committed to implementing an SMS. Level 4 status was a result of full implementation of all SMS components.

Figure 1-2 SMS Pilot Project: five levels of SMS implementation.

The final level of SMS maturity was the continuous improvement level. At this point processes were in place and their performance and effectiveness verified. The complete SA process, including continuous monitoring and the remaining features of the other SRM and SA processes were functioning. Level 4 became the major objective of any successful SMS to attain and, better yet, maintain for the life of the organization. The FAA Program Office conducted the external audits and issued a "Letter of SMS Recognition" at the successful completion of each audit.

AIRPORT SMS PILOT STUDIES

FAA Airports also initiated a number of airport SMS pilot studies to evaluate the development and implementation of SMS at airports of varying size and complexity. The pilot studies also allowed airports and the FAA to gain experience establishing airport-specific SMSs that were tailored for the individual airport.

More than 20 airports participated in the first round of the pilot studies, which were initiated in April 2007. FAA published a PowerPoint presentation summarizing the findings of the first round in October 2008. In July 2008, FAA Airports initiated a second round with nine airports to focus on the development of SMS in our nation's smaller certificated airports. FAA concluded the second round of studies and issued a final report on the combined pilot study findings.

In December 2009, FAA initiated a Part 139 SMS Implementation Study to examine how airports implement Safety Risk Management and Safety Assurance components throughout their airfield environment. Eligibility for the study was limited to airports that participated in the first or second studies. Fourteen airports participated.

14 CFR PART 5: SAFETY MANAGEMENT SYSTEMS

The long awaited **14 CFR Part 5 Safety Management Systems** rule was published on January 8, 2015. After the standard 60-day waiting period, the rule became effective March 9, 2015, requiring all 14 CFR Part 121 certificate holders to develop and implement a safety management system within their organizations. This Part 5 rule mandated that SMS be required for all Part 121 certificate holders within six months from the effective date. Additionally, the SMS Pilot Program was cancelled.

All other certificate holders who choose to implement SMS will continue to do so under the **SMS Voluntary Program (SMSVP)** using the SMSVP Standard. One of the primary differences in the SMSVP Standard and the Part 5 rule is that the dates for implementation plan submission, approval, and SMS acceptance specified in §5.1 do not apply to those that are implementing SMS under the SMSVP.

Figure 1-3   Transition from the SMS Pilot Project to the SMS Voluntary Program.

A certificate holder may develop and implement an SMS in any manner it deems appropriate. When a certificate holder requests FAA recognition of its SMS, however, an implementation plan is submitted to its Certificate Management Team (CMT) for validation against the SMSVP standard.

SMS VOLUNTARY PROGRAM

The newly created SMS Voluntary Program (SMSVP) Guide, AFS-900-002-G201, deleted the five SMS Pilot Program Levels and created three new SMSVP Levels:

1. **SMSVP Active Applicant:** the certificate holder and Certificate Management Team have committed to sufficiently support the SMS implementation and validation processes.
2. **SMSVP Active Participant:** the certificate holder officially begins and maintains its implementation efforts. The certificate holder receives an acknowledgment letter.
3. **SMSVP Active Conformance:** the Certificate Management Team and the SMS Program Office (SMSPO), located in Washington D.C., acknowledge full implementation of the certificate holder's SMS. The certificate holder receives an acknowledgment letter. The certificate holder is expected to use and continually improve its safety management processes.

THE INTERNATIONAL MANDATE FOR SMS

To fully understand international aviation operations, including aviation safety and policy, one needs to understand the ICAO and U.S. involvement with this organization.

World War II had a major impact on the technical development of aircraft, compressing 25 years of peacetime development into six years. There were many political and technical problems to resolve in support of a world at peace. Safety and regularity in air transportation made it necessary for airports to install Navigational Aids (NAVAID) and weather reporting systems. Standardization of methods for providing international services was vital to preclude unsafe conditions caused by misunderstanding or inexperience.

ICAO established standards for air navigation, air traffic control (ATC), personnel licensing, airport design, and many other important issues related to air safety. Questions concerning the commercial and legal rights of developing airlines to fly into and through the territories of another country led the United States to conduct exploratory discussions with other allied nations during early 1944. On the basis of these talks, allied and neutral states received invitations to meet in Chicago in November 1944. The outcome of this Chicago Convention was a treaty requiring ratification by 26 of the 52 states that met. By ratifying the treaty, contracting states agreed to pursue certain stated objectives, assume certain obligations, and establish the international organization that became known as ICAO.

As a charter member of ICAO, the United States has fully supported the organization's goals from its inception, especially its concerns with aviation safety. Through active support and participation in ICAO, the Federal Aviation Administration (FAA) strives to improve worldwide safety standards and procedures.

ICAO AND THE ICAO ANNEXES

Through international agreements, the ICAO Annexes contain adopted standards and recommended practices. These are referred to as the ICAO Annexes. The ICAO Annexes identify the Standards and Recommended Practices (SARPs) for all member nations to follow. Although the ICAO Annexes, or SARPS, are not mandatory, failure to follow ICAO recommendations can result in the expulsion from the ICAO as a member nation. The ramifications of losing membership to the ICAO would be economically disastrous to any nation.

The following are descriptions of the 19 ICAO Annexes:

ANNEX 1 PERSONNEL LICENSING. Provides information on licensing of flight crews, air traffic controllers, and aircraft maintenance personnel including medical standards for flight crews and air traffic controllers.

ANNEX 2 RULES OF THE AIR. Contains rules relating to conducting visual and instrument flight.

ANNEX 3 METEOROLOGICAL SERVICE FOR INTERNATIONAL AIR NAVIGATION. Provides for meteorological services for international air navigation and reporting of meteorological observations from aircraft.

ANNEX 4 AERONAUTICAL CHARTS. Contains specifications for aeronautical charts used in international aviation.

ANNEX 5 UNITS OF MEASUREMENT TO BE UWSED IN AIR AND GROUND OPERATIONS. Lists dimensional systems used in air and ground operations.

ANNEX 6 OPERATION OF AIRCRAFT. Enumerates specifications to ensure a level of safety above a prescribed minimum in similar operations throughout the world.

ANNEX 7 AIRCRAFT NATIONALITY AND REGISTRATION MARKS. Specifies requirements for registration and identification of aircraft.

ANNEX 8 AIRWORTHINESS OF AIRCRAFT. Specifies uniform procedures for certification and inspection of aircraft.

ANNEX 9 FACILITATION. Provides for the standardization and simplification of border crossing formalities.

ANNEX 10 AERONAUTICAL TELECOMMUNICATIONS. Volume 1 provides for standardizing communications equipment and systems. Volume 2 standardizes communications procedures.

ANNEX 11 AIR TRAFFIC SERVICES. Includes information on establishing and operating air traffic control (ATC), flight information, and alerting services.

ANNEX 12 SEARCH AND RESCUE. Provides information on organization and operation of facilities and services necessary for Search and Rescue (SAR).

ANNEX 13 AIRCRAFT ACCIDENT AND INCIDENT INVESTIGATION. Provides for uniformity in notifying, investigating, and reporting on aircraft accidents.

ANNEX 14 AERODROMES. Contains specifications for the design and equipment of aerodromes.

ANNEX 15 AERONAUTICAL INFORMATION SERVICES. Includes methods for collecting and disseminating aeronautical information required for flight operations.

ANNEX 16 ENVIRONMENTAL PROTECTION. Volume 1 contains specifications for aircraft noise certification, noise monitoring, and noise exposure units for land-use planning. Volume 2 contains specifications for aircraft engine emissions.

ANNEX 17 SECURITY: SAFEGUARDING INTERNATIONAL CIVIL AVIATION AGAINST ACTS OF UNLAWFUL INTERFERENCE. Specifies methods for safeguarding international civil aviation against unlawful acts of interference.

ANNEX 18 THE SAFE TRANSPORT OF DANGEROUS GOODS BY AIR. Contains specifications for labeling, packing, and shipping dangerous cargo.

ANNEX 19 SAFETY MANAGEMENT SYSTEM. Specifies the requirement for all member states to require and recognize a safety management system for most all aviation service providers.

In its March 2006 amendments to the Annexes, the ICAO established a standard for member states to mandate that each of these operators establish an SMS. Member states agreed to initiate compliance with amendments to the ICAO Annexes by January 1, 2009. The ICAO provides that each ICAO member state is the judge of whether its national SMS rules provide an acceptable level of safety. **If you are holding a certificate issued from the FAA, this guidance mandates the approval of your SMS from the FAA.**

Under ICAO Annex 19, each State shall require that the following service providers under its authority implement an SMS:

- **Approved training organizations** in accordance with Annex 1 that are exposed to safety risks related to aircraft operations during the provision of their services;
- **Operators of airplanes or helicopters** authorized to conduct international commercial air transport, in accordance with Annex 6, Part I or Part III, Section II, respectively. This includes international general aviation operators of large or turbojet airplanes in accordance with Annex 6, Part II, Section 3;
- **Approved maintenance organizations** providing services to operators of airplanes or helicopters engaged in international commercial air transport, in accordance with Annex 6, Part I or Part III, Section II, respectively;
- **Organizations responsible for the type design or manufacture of aircraft**, in accordance with Annex 8;
- **Air traffic services (ATS) providers** in accordance with Annex 11; and
- **Operators of certified aerodromes** in accordance with Annex 14.

UNITED STATES STATUARY REQUIREMENTS FOR SMS

Congress, in the Airline Safety and Federal Aviation Administration Extension Act of 2010, directed the FAA to issue a notice of proposed rulemaking within 90 days of enactment, and a final SMS rule by July 30, 2012. In addition, the National Transportation Safety Board (NTSB) recommended the FAA pursue rulemaking to require the implementation of SMS. The FAA's Air Traffic Organization (ATO) has complied with the ICAO SMS mandate for several years. Additionally, the FAA published a new 14 CFR Part 5 in January 2015, applicable to Part 121 operations. The FAA is expected to issue final guidance for airports in 2016.

The high priority objective for the FAA is to comply with ICAO standards, fully address numerous NTSB recommendations, and comply with the statutory requirements mandated by Congress.

REVIEW QUESTIONS

1. How would you define SMS?

2. Explain how the United States Congress has passed a statutory requirement for implementing an SMS Program.

3. Is the United States a member nation of the ICAO?

4. Explain how the ICAO documents the standards and recommended practices to its member nations?

5. Explain how the ICAO has mandated the use of SMS.

6. What is the difference between the FAA's SMS Pilot Program and 14 CFR Part 5.

7. What is the difference between the FAA's SMS Pilot Program and the SMSVP?

8. For effective safety reporting people must be willing to report their errors and experiences.
    a. True.
    b. False.

9. Under a safety management system, what item would be found under safety policy?
    a. Procedures.
    b. Organization.
    c. Training.
    d. Both a and b.

10. Under a safety management system, what item would be found under safety risk management?
    a. Risk mitigation.
    b. Internal evaluation program.
    c. Communication.
    d. None of the above.

11. What ICAO Annex is directly related to SMS?
    a. Annex 1.
    b. Annex 6.
    c. Annex 17.
    d. Annex 19.

12. Which is a subculture of a safety culture?
    a. Indicating culture.
    b. Informed culture.
    c. Identifying culture.
    d. Educated culture.

13. The FAA established the SMS Voluntary Program prior to the FAA SMS Pilot Programs.
    a. True.
    b. False.

14. SMS is a structured process that obligates organizations to:
    a. Manage safety with higher priority than other core business processes are managed.
    b. Manage safety with the same level of priority that other core business processes are managed.
    c. Identify hazards and remove all risk from the organization.
    d. Manage safety with safety objectives that include minimizing expenses.

15. What are the four components of safety management systems?
    a. Safety policy, safety regulations, safety procedures, and safety culture.
    b. Safety policy, safety risk management, hazard identification, and safety assurance.
    c. Safety policy, safety risk management, safety assurance, and safety promotion.
    d. Safety policy, safety promotion, safety culture, and safety protection.

CHAPTER 2
Understanding the Components of an SMS

OBJECTIVES

- To individually analyze the four components of a safety management system.
- To synthesize and evaluate how the four components harmonize to create an effective safety management system.

KEY TERMS

- Accountability
- Accountable Executive
- Audit
- Aviation Safety Action Program (ASAP)
- Certificate Holder
- Data Collection Tool
- Evaluation
- FAA Certificate Management Team (CMT)
- Investigation
- Key Safety Management Personnel
- Output
- Process Manager
- Responsibility
- Risk Matrix
- Safety Objective
- Safety Promotion
- Safety Risk Assessment (SRA) Tool
- Substitute Risk

INTRODUCTION

A safety management system consists of four components: **safety policy**, **safety risk management**, **safety assurance**, and **safety promotion**. The challenge in discussing any one component of SMS is each component only works when it is used in harmony with the other three components. Similarly, when operating an automobile, anyone who has driven a car knows all four wheels must be turning simultaneously, in harmony for the automobile to function properly. Someone who hasn't driven a car before might consider a two-wheel car is all that is really needed but once you start going down the road, it won't take long to recognize something is amiss.

Similarly, as you begin learning about or using SMS, the temptation is to focus on certain components and not recognize the importance or harmony of all the components within the system. Knowing this relationship is critical as you develop and maintain a safety management system within your organization. With this in mind, we will begin this chapter by separating each component and describing its function. We will summarize, by pulling the four components together to understand how they work in harmony.

Figure 2-1 The four components of SMS. *(FAA.)*

SAFETY POLICY COMPONENT

Safety policy is not the most exciting of the four components. The word "policy" may trigger a Pavlovian response of a long sigh and a roll of the eyes. Policy scares us; it brings images of large mounds of paperwork, standard operating procedures, do's and don'ts, and details that hinder us from getting the job done quickly. Fortunately, the purpose of safety policy has far more strategic goals that lay the groundwork for the other three components.

> **Safety policy** means the certificate holder's documented commitment to safety, which defines its safety objectives and the accountabilities and responsibilities of its employees in regards to safety.

When developing a safety management system, Safety Policy is where the work normally begins. The certificate holder (aviation service provider) must start developing the policy framework to enable development of SMS to continue. This framework requires two parts:

1. Most importantly it provides the purpose of establishing performance in relation to the safety objectives and accountabilities and responsibilities of all it's employees in regard to safety; and

2. An actual written safety policy, signed by the accountable executive, that reflects documentation that all its components are in operation.

Establishing Safety Policy Performance

Safety policy begins with the **accountable executive**. The accountable executive is the individual who a certificate holder or service provider designates as the individual who has final authority over operations authorized under its organization and is ultimately responsible for their company's safety performance. Having commitment from the person who controls financial and human resources is critical to the long-term success of SMS. To begin the commitment, the accountable executive, designated because of his authority within the organization, signs safety policy committing to provide adequate resources for SMS development, implement SMS in all relevant areas of their organization, and ensure ongoing conformance to either 14 CFR Part 5 of the federal regulations or SMSVP Standard.

> The **accountable executive** has the ultimate responsibility for safety management within the organization.

Understandably, if key leadership is not behind the development and maintaining of SMS it will be short lived and soon revert back to a safety program. Safety policy requires not only the definition of the accountable executive but also all members of management in regard to developing, implementing, and maintaining SMS processes within their area of responsibility. These individuals are often called **key safety management personnel**.

> **Key safety management personnel** are the individuals who are responsible for identifying hazards, conducting risk assessments, and developing risk controls for their areas of responsibility. They have the technical expertise and are the ones responsible for implementation and operation of risk controls (often in the form of operational procedures, specified tools, training, communication, etc.)

These key safety management personnel reflect not only the vice president of aviation safety, but all the vice presidents and directors of the organization, or whoever makes decisions for the organization. This is a critical departure from a safety program in that leadership personnel are not only responsible and evaluated on production, but also **responsible** and evaluated on the protection. Their job descriptions will establish who is responsible for what areas of safety management as well as who has the **authority** to accept risk within the organization. As your SMS matures, all employees will have safety related performance standards within their job function in which they will be evaluated on to determine satisfactory job performance. These safety performance standards must be specific, measurable, and attainable for each employee. For example, at the end of the year evaluation, lower management may be evaluated on whether they attended all the monthly safety meetings.

Accountability refers to active management and line employee involvement and action in managing and maintaining safety performance. A certificate holder defines accountability by ensuring that each of its management and line employees is aware of his or her specific role within SMS and actively participates in carrying out his or her SMS-related duties. Once the accountabilities for these employees have been defined, Subpart D (Safety Promotion) requires that these accountabilities be communicated throughout the organization.

Safety objectives must be established with the participation of all decision makers recognized as key safety management personnel. These objectives will establish the purpose for all the processes, procedures, and changes made related to safety. They help us measure success and determine what we will measure. In Chapter 10 we will discuss more regarding how to practically develop a safety policy and safety objectives.

Lastly, an **emergency response plan** must be established. This plan should provide procedures for management decision-making and action in an emergency. This should include a line of succession of management authority sufficient to respond to emergencies. Coordination of your emergency response plans with the emergency response plans of other organizations might include first responders to accidents or incidents, airport authorities, and hazardous materials (hazmat) authorities. The plan might also address how you return or transition to normal operations after the emergency condition subsides. Many organizations already have emergency response plans that may be used to fulfill this requirement.

These parts make up a larger whole, recognizing employees don't just know these things through corporate knowledge. It will never paint a clear picture of what is transpiring and what is expected. Therefore, safety policy is required and in general principle, what organizations do, and what they expect from their employees must be documented and trained.

Written Safety Policy

14 CFR Part 5, Subpart B—Safety Policy requires the following written documentation. Listed below reflect the federal regulations regarding what is needed to be recognized by the FAA to have a safety management system at your organization.

§5.21 Safety policy.

(a) The certificate holder must have a safety policy that includes at least the following:

(1) The safety objectives of the certificate holder.

(2) A commitment of the certificate holder to fulfill the organization's safety objectives.

(3) A clear statement about the provision of the necessary resources for the implementation of the SMS.

(4) A safety reporting policy that defines requirements for employee reporting of safety hazards or issues.

(5) A policy that defines unacceptable behavior and conditions for disciplinary action.

(6) An emergency response plan that provides for the safe transition from normal to emergency operations in accordance with the requirements of §5.27.

(b) The safety policy must be signed by the accountable executive described in §5.25.

(c) The safety policy must be documented and communicated throughout the certificate holder's organization.

(d) The safety policy must be regularly reviewed by the accountable executive to ensure it remains relevant and appropriate to the certificate holder.

§5.23 Safety accountability and authority.

(a) The certificate holder must define accountability for safety within the organization's safety policy for the following individuals:

(1) Accountable executive, as described in §5.25.

(2) All members of management in regard to developing, implementing, and maintaining SMS processes within their area of responsibility, including, but not limited to:

(i) Hazard identification and safety risk assessment.

(ii) Assuring the effectiveness of safety risk controls.

(iii) Promoting safety as required in subpart E of this Standard.

(iv) Advising the accountable executive on the performance of the SMS and on any need for improvement.

(3) Employees relative to the certificate holder's safety performance.

(b) The certificate holder must identify the levels of management with the authority to make decisions regarding safety risk acceptance.

§5.25 Designation and responsibilities of required safety management personnel.

(a) *Designation of the accountable executive.* The certificate holder must identify an accountable executive who, irrespective of other functions, satisfies the following:

(1) Is the final authority over operations authorized to be conducted under the certificate holder's certificate(s).

(2) Controls the financial resources required for the operations to be conducted under the certificate holder's certificate(s).

(3) Controls the human resources required for the operations authorized to be conducted under the certificate holder's certificate(s).

(4) Retains ultimate responsibility for the safety performance of the operations conducted under the certificate holder's certificate.

(b) *Responsibilities of the accountable executive.* The accountable executive must accomplish the following:

(1) Ensure that the SMS is properly implemented and performing in all areas of the certificate holder's organization.

(2) Develop and sign the safety policy of the certificate holder.

(3) Communicate the safety policy throughout the certificate holder's organization.

(4) Regularly review the certificate holder's safety policy to ensure it remains relevant and appropriate to the certificate holder.

(5) Regularly review the safety performance of the certificate holder's organization and direct actions necessary to address substandard safety performance in accordance with §5.75.

(Continued)

(c) Designation of management personnel. The accountable executive must designate sufficient management personnel who, on behalf of the accountable executive are responsible for the following:

(1) Coordinate implementation, maintenance, and integration of the SMS throughout the certificate holder's organization.

(2) Facilitate hazard identification and safety risk analysis.

(3) Monitor the effectiveness of safety risk controls.

(4) Ensure safety promotion throughout the certificate holder's organization as required in subpart E of this Standard.

(5) Regularly report to the accountable executive on the performance of the SMS and on any need for improvement.

§5.27 Coordination of emergency response planning.

Where emergency response procedures are necessary, the certificate holder must develop and the accountable executive must approve as part of the safety policy, an emergency response plan that addresses at least the following:

(a) Delegation of emergency authority throughout the certificate holder's organization;

(b) Assignment of employee responsibilities during the emergency; and

(c) Coordination of the certificate holder's emergency response plans with the emergency response plans of other organizations it must interface with during the provision of its services.

SAFETY RISK MANAGEMENT COMPONENT

The SRM component is the component that is most visibly seen, it is the one on the surface makes the most sense. It is also takes the most work and practice to do properly. You can read and study all the aspects of SRM, but until you actually work through actual risk management exercises where you identify hazards, determine and mitigate risk, you will have only part of the picture.

Safety risk management is a process within the SMS composed of describing the system, identifying the hazards, and analyzing, assessing and controlling risk.

As stated in the definition, SRM is a process. A process is a systematic series of actions directed to some end; which is precisely what SRM is trying to accomplish. It is a set of actions with the end purpose to reduce risk to the lowest practical level. As discussed in earlier terms reducing risk to the lowest practical does not necessarily mean eliminating risk. It is bringing it down to the lowest level that in turn, will still allow the mission to be accomplished and the risk low enough in which management will be willing to accept the remaining risk.

SRM is a formal system for identifying and mitigating risk. There are five processes necessary to control and mitigate risk. These processes are featured in Figure 2-2, safety risk management processes and regulatory requirements, and are discussed in more detail below. These are:

1. System description and analysis.
2. Hazard identification.
3. Safety risk analysis.
4. Safety risk assessment.
5. Safety risk controls.

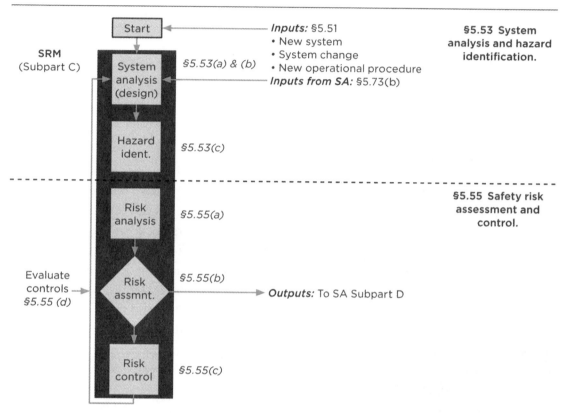

Figure 2-2   SRM processes and regulatory requirements.

System Description and Analysis

To know when SRM process may be required, we must be able to define the system as described in Chapter 3. The system could be people, departments, hardware, software, information, procedures, facilities, services, and other support facets that are directly related to the organization's aviation safety activities. Examples from AC 120-92B of broad-based systems and subsystems could include:

1. Flight operations
   a. Aircraft type
   b. Fixed wing
   c. Rotocraft
   d. Unmanned
   e. Crew scheduling

2. Operational control
3. Records
4. Standards
5. Maintenance and inspection
    a. Quality assurance
    b. Maintenance control
6. Cabin safety
7. Ground handling and servicing
    a. Deicing
    b. Fueling
8. Cargo handling
9. Training
    a. Training curriculum
    b. Hazmat training

If we do not properly identify the system, it is a like a doctor making a misdiagnosis. The misdiagnosis will result in an improper treatment plan, which in turn can result in more damage than good being done. This is the same thing when your organization misdiagnosis the system being analyzed. We apply incorrect risk mitigation strategies and more importantly inaccurately identify the hazards and hazard consequences associated with those hazards.

An effective method to ensure you have not misdiagnosed your system is to invite the appropriate people to the table. Experience matters, those who understand the entire organization are often the individuals who will be able to identify the entire system and who will be affected.

The SRM process is triggered when proposed new systems or changes to systems are being considered. Examples for different operations could include:

- Changes to your operation could include the addition of new routes;
- Opening or closing of line stations;
- Adding or changing contractual arrangements for services;
- Additions of new fleets or major modifications of existing fleets;
- Addition of different types of operations such as extended operations (ETOPS);
- New equipment to your flight training aircraft such as angle of attack indicators;
- A new hangar or other facility;
- A new position needed for growth;
- Unplanned or rapid growth; or
- Any one of many different types of operations.

The list is literally unlimited; change must be managed.

The SRM process is not triggered solely by major changes to a system; it is triggered by any revision of an existing system, development of operational procedures, and identification of hazards or ineffective risk controls through the safety assurance processes that will be described under safety assurance component. The level of SRM documentation needed for smaller changes to a system may be significantly smaller than for major changes. It should be noted, as per AC 120-92B states, "It is not the intent of Part 5 to require the application of SRM processes and procedures to activities that are not related to aviation operations."

Triggers to conduct SRM Process are:

- New System.
- Change to Existing System.
- Development of Operational Procedures.
- Identification of hazards or ineffective risk controls.

The outputs of managing risk could be recorded in a simple recording medium such as a worksheet or a notebook, common desktop software, or a web-based application tool (WBAT). Later in this chapter we will provide an example of a worksheet that could be used as paper records or converted to a variety of software applications, including desktop spreadsheets or WBATs. This recording medium is called a **safety risk assessment (SRA) tool**.

Systems analysis is the primary means of proactively identifying and addressing potential problems before the new or revised systems or procedures are put into place. The system analysis should explain the functions and interactions among the hardware, software, people, and environment that make up the system in sufficient detail to identify hazards and perform risk analyses. The process is started by describing the system (this can be as simple as flowcharting the system or writing a short narrative—see Figure 2-3 on the next page).

An example of system description and analysis might involve the need in your operations for a new aircraft (or fleet of aircraft) to meet your company or corporate goals. Several of your organizational systems would be affected: flight operations, maintenance, station or ground. As part of your examination of the flight operations system, you need to consider changes to pilot qualifications, pilot and mechanic training, scheduling, crew rest, union participation, and other areas. This is a process normally done as part of your business activities.

Your system analysis should identify and consider activities and resources necessary for the system to function. For example, in the scenario of adding aircraft to your fleet, you would identify the pilot training system as one of the affected systems, in particular the activities and resources necessary for pilot training to fly the additional aircraft. These may include simulators, training curriculum, training aids, and instructors.

The system description and analysis process frequently includes appropriate representatives from management, safety staff, subject matter experts (SME), employees, and representation groups (such as unions) formed into workgroups such as safety committees, safety roundtables, safety action groups, or similar titles. Since many, if not most, system changes involve allocation of resources, the accountable executive or other managers with the authority to commit resources should be included in the process.

1. Regardless of the type of organization, aviation or non-aviation related, what are some recent changes made?
   a. How did they or did they not take into account the entire system?
   b. Who were the "appropriate" people invited to plan for the change? Were people left out? Why?
2. What are some different systems within your organization?
3. How does your organization determine who should be involved in managing risk within your organization?

Hazard Identification

The hazard identification process flows from the system analysis. When using a safety risk assessment tool to identify hazards we may ask: What will keep me up at night? What really worries me? What would worry the accountable executive? What could go wrong with our processes? Under typical or abnormal operational conditions, that could cause an accident? Generally for our organization, we also need to ask ourselves what current processes do we have in place to help us answer these questions and what processes should we have in place in the future.

Most often the same individuals or groups conducting the system analysis process (safety committees, safety roundtables, etc.) would conduct hazard identification using a safety risk assessment tool. Experience, FAA requirements, manufacturer's technical data, and knowledge of your operations are used to identify hazards to your operation. For example, as depicted in Figure 2-3, UND Aerospace was replacing their fleet of aircraft at their Phoenix satellite location, with an entire different training aircraft. Hazards could include the effectiveness of new procedure training, employees missing training, failing understand newly published procedures, supervisors failing to monitor the new procedures, etc. While identification of every conceivable hazard is unlikely, you are expected to exercise due diligence in identifying foreseeable hazards that could lead to an aircraft accident. Foreseeable, suggests a emphasis on proactively identifying hazards before harm or damage is done to an organization or individual.

For each identified hazard, continue to define the potential for injury and damage that may result from a mishap related to operating while exposed to the hazard. Ask what is it about the factors analyzed, individually or in combination, which could result in an accident.

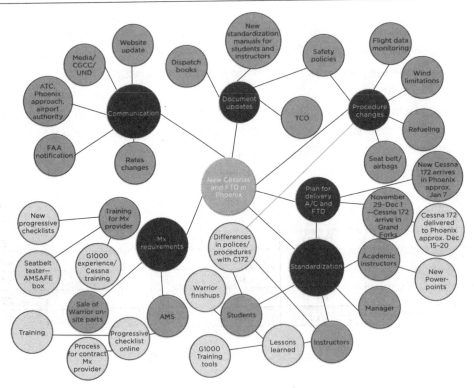

Figure 2-3   SRM processes and regulatory requirements. *(UND Aerospace)*

Risk Analysis

Once we have identified the hazards and hazard consequences we need to make sense of all the data and then assess what level of risk exists within the hazard consequence. As we analyze the data we ask questions, such as, "what does all this information really tell us?" Sometimes it is analyzing the hazards and hazard consequences we have already recorded in a safety meeting, other times it is looking at the data gained through cockpit voice records and recorded flight data. For example, your flight training organization records all the flight parameters on a given flight including exceedances of aircraft limitations. You identify the hazard of pilots exceeding aircraft limitation, but now you need to make sense of the data. Are pilots really exceeding limitations or are the limitations set too high? Are there other factors contributing? How often is it happening, one out of every 3,000 flights? Regardless, the intent is to analyze the identified hazards and hazard consequences as it relates with the overall system.

This is one reason why the system analysis is an essential foundational step in risk management. If risk analysis is not based upon a thorough understanding of the system, you may miss important details that could cause the system to fail. The knowledge gained in the system analysis and subsequent risk analysis will later be used to develop a mitigating strategy. Risk controls will target the conditions that we think will cause an accident or affect its severity or likelihood.

The risk analysis also needs to consider the basis for the estimates of severity and likelihood. Have you changed equipment that your employees must use, the procedures for using it, the layout of the facility, etc., in ways that could increase the likelihood of errors resulting in an accident? For example, if, in the process of a merger, flight deck procedures from one of the partner airlines become the standard across the merged carrier and if the change in procedures has been identified as a hazard, what is it about the new procedures that would lead to errors?

In order to determine potential for injury and damage, you need to define the likelihood of the occurrence of an accident and severity of the injury or damage that may result from the all the hazard consequences you have identified. It is important to remember that the likelihood and severity do not refer to the hazard but of a potential occurrence (hazard consequence) related to the hazard. When conducting a risk assessment, a tool to assist you in making the decision is called a **risk matrix** (see Figure 2-4).

Likelihood		Severity				
		Catastrophic	Severe	Major	Minor	Negligible
		A	B	C	D	E
Frequent	5	5A	5B	5C	5D	5E
Occasional	4	4A	4B	4C	4D	4E
Remote	3	3A	3B	3C	3D	3E
Improbable	2	2A	2B	3C	3D	3E
Extremely Improbable	1	1A	1B	1C	1D	1E

Figure 2-4 Risk matrix.

As discussed in other chapters, two general methods are accepted in regard to assigning a likelihood and severity.

1. Mathematical probabilities—used most often by airlines and engineers
2. Subject matter experts—successfully gathering the subject matter experts who understand the system affected as well as the technical aspects to determine the likelihood and severity. This method can be very effective, but if SMS is not organizationally accepted or the management involved is unfamiliar with the process, the results can be suspect. Risk assessment in operational contexts are often based on expertise and expert judgment, but they should also use data from the carrier or service provider's own experience or those of others in the industry where available. Review of accident statistics, failure data, error data (runway incursion reports or information from the National Aeronautics and Space Administration's (NASA) Aviation Safety Reporting System (ASRS) or equipment reliability data may help in determining likelihood.)

The type of consequence (error, failure, accident, or incident) that is envisioned normally drives the estimate of severity. For example, if the hazard could result in controlled flight into terrain (CFIT), the severity of this outcome is normally major, if not catastrophic. Conversely, tire failures, while potentially leading to a fatal accident, more often lead only to aircraft damage. As an organization you will have to develop a risk matrix tool to help you assess or decide what the severity and likelihood is for any given hazard consequence identified.

Even where the best estimate has to be based on reasonable expert judgment, effective risk management can be accomplished by applying a disciplined analysis using a risk assessment and risk matrix tool developed by your organization.

Risk Assessment

Once the risk is analyzed, you must assess or decide whether the risk is acceptable. A risk matrix tool provides you with a way to integrate the effect of severity of the outcome and the probability of occurrence, which enables you to:

- Frame thought process for analysis;
- Assess risks;
- Compare potential effectiveness of proposed risk controls; and
- Prioritize risks where multiple risks are present.

If a risk matrix is used, the carrier should develop criteria for severity and likelihood that are appropriate for their type of operations and their operational scenario. For example, severity levels are sometimes defined in terms of a dollar value of potential damage. In this case, different types of airplanes operated and their relative values would dictate different scales between carriers. Likewise, the method that the carrier uses to estimate likelihood will have an effect on how likelihood scales are defined. If the carrier prefers to use quantitative estimates (probability), the scales would be different than one that prefers to use qualitative estimates. Additional details regarding developing a risk matrix tool can be found in Chapter 11.

As shown in Figure 2-4, the risk matrix tool will be used a second time to assess the new level of risk after proposed risk controls are established. This is helpful part of the process and discussion as your organization determines if they believe the risk controls established will reduce the risk to the lowest practical and acceptable level.

Assessing risk using a standard risk matrix tool also assists in the communication process and to properly prioritize risks when multiple hazard consequences are present. This allows the work to be done first on the high risk, high priority items instead of spending too much time on the hazard with hazard consequences that are not as worrisome to the organization.

Risk assessment is based on judgment, experience, and input from data collection tools (DCT) and previous processes. If the risk is at the lowest practical level or the change you are proposing has no new hazards, you may elect to be done with the SRM process. If this is the case, then the system may be placed into operation and monitored in the safety assurance process. If you changed the order of a procedure but no content, you might determine that no new hazards exist. Regardless of no new hazard exists from the change, you must manage that change well, or you will create new hazards quickly. For example, you change the order of a procedure from Step 1, Step 2, Step 3 to Step 1 then Step 3, and finally Step 2. If you assess the risk and determine the re-ordering of the steps will create no new hazard. You must manage that change. Where will the change be written (safety policy), who will be responsible to implement the change (safety policy), how will we tell and train those affected (safety promotion), how will ensure it does what we expected the change to do (safety assurance)?

In contrast, if you do decide the risk is not acceptable, or you feel it is not at the lowest practical level, you will need to follow the next step in SRM, which is developing risk controls.

Example of Likelihood and Severity Criteria

Likelihood of Occurrence.

    5 Frequent—Likely to occur many times (likely to occur within 30 days).

    4 Occasional—Likely to occur sometimes (probability will occur within 6 months to 1 year).

    3 Remote—Unlikely to occur, but possible (has occurred rarely).

    2 Improbable—Very unlikely to occur (not known to have occurred).

    1 Extremely Improbable—Almost inconceivable that the event will occur.

Severity of Occurrence. Assuming that this event indeed occurs, what would be the likely result?

    A Catastrophic—A death or multiple deaths and/or equipment destroyed.

    B Severe—Serious injury and/or major equipment damage.

    C Major—Injury to persons and/or damage to equipment.

    D Minor—Incident (minor damage to equipment and no injury to persons).

    E Negligible—Little consequences.

Risk Acceptance

Risk assessments must also involve the levels of management with the authority to make risk decisions, deciding what is or is not an acceptable risk for the systems within their area of operational responsibility. SMS requires this be documented and those who accepted risk are trained. The higher the risk, the higher the level of management is needed to accept that risk. Again, this is something that your organization must develop to meet your organizations structure and needs.

To make this process easier you will note the risk matrix tool in Figure 2-4 is color-coded. This not only helps to categorize the significance of the risk but to signify who can accept the risk. For example, if dispatching a flight presents a low or medium risk it might require the chief pilot or director of operations to approve or authorize the flight. Low risk or risk deemed acceptable, may be color coded as green, while medium risk, meaning the risk is acceptable with mitigation, is yellow. For large scale operational decisions, when risk has been assessed as high, or unacceptable, it is color coded red, the accountable executive may be the only appropriate person to make these risk acceptance decisions. Thus, the person responsible for making these risk acceptance decisions will depend on the scope of the proposed change to the operation and the level of overall risk presented to the organization.

Listed below in Figure 2-5 is a an example of how one flight training organization color codes their risk levels to identify the significance of the risk and to determine who can accept that risk.

RISK ACCEPTANCE

The acceptability of risk is evaluated using the risk matrix on the previous page. The matrix is color coded; unacceptable (red), acceptable (green), and acceptable with mitigation (yellow). The acceptance criteria and designation of authority and responsibility for risk management decision making is as follows:

The accountable executive has final authority over all operations and can accept or deny risk at any level.

(1) Unacceptable (Red)

The risk is assessed as unacceptable and further work is required to design an intervention to eliminate the hazard or to control the factors that lead to higher risk likelihood or severity. The activity that causes exposure to this risk is ceased at once, if it is ongoing. If the activity is not ongoing, it shall not be commenced, until mitigation strategies can reduce the severity and/or likelihood of the event occurring to acceptable levels. The accountable executive has final authority over all operations and can accept or deny risk at any level.

(2) Acceptable with Mitigation by Proper Acceptance Authorities (Yellow)

The risk may be accepted under defined conditions of mitigation. The acceptance is by the appropriate director or the accountable executive, following mitigation as appropriate. In addition to the appropriate operational director, the director of aviation safety must concur in the acceptance. These situations require continued special emphasis in the safety assurance (SA) function to ensure the continued effectiveness and compliance of the risk mitigation (controls and defenses, usually in the form of new policies and procedures) by spot checks and audits.

(3) Acceptable (Green)

Where the assessed risk has been reduced from red to green or yellow to green, the risk mitigations will be accepted by the appropriate director or the accountable executive. In addition, the director of aviation safety must concur in the acceptance.

Figure 2-5 Risk acceptance example.

Where the assessed risk is in the green, after factoring in existing procedures and controls, the appropriate key safety management personnel or process managers may accept the risk mitigations without further action. Safety risk management should always be used to reduce risk to as low as practicable regardless of whether or not the assessment shows that it can be accepted as is. This is a fundamental principle of continuous improvement.

Risk Controls

After hazards and associated risk are fully understood, risk controls must be designed for risks that the carrier deems unacceptable. This is accomplished using their risk assessment process, such as a SRA tool, as specified in §5.55(b). Examples of risk controls are explain in Chapter 11 but include new processes, equipment, training, new supervisory controls, new equipment or hardware, new software, changes to staffing arrangements, or any of a number of other system changes. In short, anything that would lessen the likelihood or severity of a potential hazard consequence.

Once the proposed controls have been decided, remember the risk must be assessed to determine if the level of risk is lowered to an acceptable level and the proposed control does not introduce unintended consequences or new hazards. This is commonly referred to as substitute risk. Section 5.55(d) of the Code of Federal regulations requires you to evaluate whether the risk will be acceptable applied in a system operational context.

Scenario for Group Discussion: Applying the SRM Process

Read the following scenario; assume this organization had a "strong" traditional safety program. Discuss the questions listed after the scenario.

An airline receives notification that the instrument landing system (ILS) at one of its destination airports is scheduled to be out of service for a period of time and, therefore, the flights will have to use a less familiar non-precision approach (NPA) procedure. The flight crews express concern about the procedure and the differences between fleets that will affect operational procedures and the currency of training. Under §5.51, this triggers the SRM processes and procedures.

Pursuant to §5.53(a) and (b), the SRM process starts with an analysis of the systems involved with the addition of a new procedure for the aircraft fleet. This would include analysis of the flight operations and training program. In this case, the company has two fleets of airplanes that are operated to the destination airport in question. The equipment, and therefore the procedures and flight crew training in these two fleets, differs. By analyzing the system, they identify a hazard (per §5.53(c)). Pilots are not properly trained to fly the procedures that are required for that specific airport.

To analyze the risk associated with this hazard (as required by §5.55(a)), a team is assembled with representatives from management, the training and standardization organization, the flight safety organization, the organization responsible for dispatch and operational control, and the pilot's union. Representatives of the **FAA Certificate Management Team (CMT)** also attend the work team's meetings. The risk analysis consists of documented discussions among the assembled SMEs.

As a result of the risk analysis, a risk assessment is done (per §5.55(b)). As part of the assessment, it is determined that the existing operational procedures and training used by one of the fleets are not acceptable for safe operation into the airport without revision and additional training. A management decision is made that the procedures and associated training currently in use will not provide an acceptable level of safety (ALoS) for operations into the airport.

After determining that the risk is unacceptable, risk controls are designed under §5.55(c). The risk control in this case consists of a change to operational procedures, including programming of the flight management system (FMS) and associated approach procedure training for flight crewmembers. Prior to implementation of the risk controls, the company checks the resulting procedures in flight simulators to ensure that the risk control will mitigate risk to an acceptable level, as required by §5.55(d).

Per §5.23(b), the solution is submitted to the fleet chief pilots for management review and acceptance of the mitigated risk. Division-level management and top management are briefed on the operation at regular management reviews. Once all concerned approve the controls they are implemented operationally. Per the safety assurance requirements, these controls are specifically targeted for performance monitoring by the operational control/dispatch organization and the flight operations organization line checks.

Answer and discuss the following questions.

1. What is the system in this scenario?
2. How was the hazard identified? What processes within your organization would allow someone to express his or her concerns?
3. What did they do to analyze the hazards?
4. How was the risk assessed?
5. By implementing controls, did this reduce the likelihood or the severity of the risk? Explain your answer.

14 CFR Part 5, Subpart C—Safety Risk Management

§5.51 Applicability.
A certificate holder must apply safety risk management to the following:
(a) Implementation of new systems.
(b) Revision of existing systems.
(c) Development of operational procedures.
(d) Identification of hazards or ineffective risk controls through the safety assurance processes in subpart D of this part.

§5.53 System analysis and hazard identification.
(a) When applying safety risk management, the certificate holder must analyze the systems identified in §5.51. Those system analyses must be used to identify hazards under paragraph (c) of this section, and in developing and implementing risk controls related to the system under §5.55(c).
(b) In conducting the system analysis, the following information must be considered:
(1) Function and purpose of the system.
(2) The system's operating environment.
(3) An outline of the system's processes and procedures.
(4) The personnel, equipment, and facilities necessary for operation of the system.
(c) The certificate holder must develop and maintain processes to identify hazards within the context of the system analysis.

§5.55 Safety risk assessment and control.

  (a) The certificate holder must develop and maintain processes to analyze safety risk associated with the hazards identified in §5.53(c).

  (b) The certificate holder must define a process for conducting risk assessment that allows for the determination of acceptable safety risk.

  (c) The certificate holder must develop and maintain processes to develop safety risk controls that are necessary as a result of the safety risk assessment process under paragraph (b) of this section.

  (d) The certificate holder must evaluate whether the risk will be acceptable with the proposed safety risk control applied, before the safety risk control is implemented.

SAFETY ASSURANCE COMPONENT

Under a safety program you most likely will not see the safety assurance component as it related to safety risk management in the previous section. Normally, risks are determined, controls are implemented, and within a safety program framework, the problem is fixed.

Safety assurance continues after SRM and evaluates the continued effectiveness of implemented risk control strategies and supports the identification of new hazards. SMS processes are in place that systematically provides confidence that organizational outputs meet or exceed safety objectives and the intended purpose of the change.

Safety assurance means processes within the SMS that function systematically to ensure the performance and effectiveness of safety risk controls, as well as assuring that the organization meets or exceeds its safety objectives through the collection, analysis, and assessment of information.

SA, as SRM, adheres to a formal system for processes necessary to assure safety. These processes are featured in Figure 2-6 on the next page.

System Operation/Monitoring

Monitoring operational processes is what supervisors do on a day-to-day basis (e.g., direct supervision of employee activities, monitoring of pilot currency, and monitoring minimum equipment list (MEL) status). Monitoring also involves reviewing data that is collected for operational purposes to look for anything of safety significance (e.g., duty logs, crew reports, work cards, process sheets, and reports from the employee safety feedback system). This may include monitoring products and services from outside sources that are used in the certificate holder's operations.

Monitoring of the operational environment involves practices that are similar to those of monitoring operational processes. The context for monitoring the operational environment of a system is developed from the system analysis that is conducted under SRM. Once the scope of the operational environment is defined under SRM, the operational environment must be monitored to assess impacts on aviation safety. For example, increases in the price of fuel may require airlines to change their scheduling, routes, and aircraft utilization.

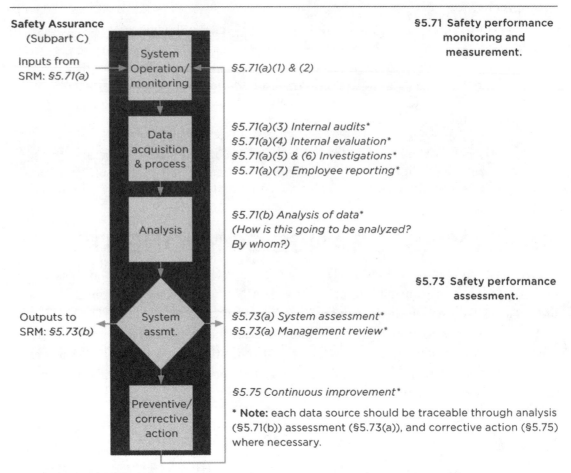

Safety Assurance
(Subpart C)

Inputs from
SRM: *§5.71(a)*

**§5.71 Safety performance
monitoring and
measurement.**

§5.71(a)(1) & (2)

§5.71(a)(3) Internal audits*
§5.71(a)(4) Internal evaluation*
§5.71(a)(5) & (6) Investigations*
§5.71(a)(7) Employee reporting*

§5.71(b) Analysis of data*
*(How is this going to be analyzed?
By whom?)*

**§5.73 Safety performance
assessment.**

Outputs to
SRM: *§5.73(b)*

§5.73(a) System assessment*
§5.73(a) Management review*

§5.75 Continuous improvement*

* Note: each data source should be traceable through analysis
(§5.71(b)) assessment (§5.73(a)), and corrective action (§5.75)
where necessary.

Figure 2-6   Safety assurance processes and regulatory requirements.

Data Acquisition & Processes

Investigations should be treated as an opportunity for organizational learning to prevent a repeat of errors and/or change company processes so that mistakes do not recur. Investigations should focus on what went wrong rather than who caused the error and emphasize improvement of safety performance. The organization should include investigation data, if available, from outside sources such as FAA or NTSB investigations and may, where appropriate, participate as a party to official investigations. The results of investigations can be recorded as well.

The investigation should reveal information that, when utilized correctly, will concentrate on objective facts to identify system deficiencies, help prevent future recurrences, and improve system reliability. It is not as important to identify "who did it" as it is for you to learn why it happened. Within this process, it is important to distinguish between error and intentional/willful noncompliant actions. Investigations of reports regarding potential noncompliance with regulatory standards or of inadequate safety risk controls established by the certificate holder should be mitigated through the SRM process. Instances of noncompliance with an FAA regulation may be reported through the Voluntary Disclosure Reporting Program (VDRP), where applicable. For instances involving individual employee noncompliance with FAA regulations, these employees may use an **Aviation Safety Action Program (ASAP)**, if one is available.

Audits are a means of collecting data to confirm whether or not actual practices are being followed within a department. Audits should typically involve the operational management responsible for the system(s) being audited. Procedures for auditing should describe your audit process, criteria, scope, frequency, method for selecting auditors, and methods of documentation and recordkeeping. Audit planning should take into account the safety criticality of the processes to be audited and the results of previous audits. Auditors should not audit their own work, but may audit the work of others around them in the same department. Audit procedures should include the responsibilities and expectations for planning, conducting, reporting results of audits, maintaining records of audit results, and processes for auditing contractors and vendors, as necessary. The results of audits will depend on the size of your organization. They can be recorded in paper format such as a common logbook-style binder, or in electronic media such as a desktop spreadsheet program or other software program available.

The **evaluation** process builds on the concepts of audit and inspection. An evaluation is an internal oversight tool that provides the accountable executive with a snapshot of the safety performance of the carrier's operational processes and systems, as well as SMS processes. The evaluation should include all available data about the organization, including information from the audits conducted by the operational management.

Conducting evaluations at planned intervals will help the carrier or service provider's management to determine if its safety management methods and practices are meeting safety objectives and expectations set out in its safety policy. Evaluation planning should take into account the safety criticality of the processes that are being evaluated and the results from previous evaluations. The scope, content, and frequency of evaluations should be based on the decision maker's need for information to assess the health of operational processes and the SMS. The decision maker will also need to define criteria for selecting evaluators.

The carrier or service provider's workforce is an important information source that should be included in the data-gathering process. Audits and evaluations bring decision makers important information and are essential to a proactive SA approach. These tools can be limited, however, by the scope and content of their design. Front-line employees may observe aspects of the operation or environment that were not expected and were not included in audit or evaluation protocols. The employee reporting system can fill in important gaps in the company's data collection process.

To be effective, the organization needs to establish and maintain an environment where employees feel comfortable reporting hazards, issues, and concerns, as well as occurrences and incidents. They should feel comfortable proposing safety solutions and improvements, as well. The accountable executive and management team need to encourage employees to report safety issues without fear of reprisals from management. Policies that assure employees fair treatment and clear standards of behavior are an essential part of the reporting process.

A key aspect of a confidential reporting system is that it is confidential. Therefore, you must define methods for employee reporting and de-identification of sources without losing essential information. As you develop and employ the confidential reporting procedures and include this invaluable input in safety decision-making, employees will begin to trust the system to work toward elimination of systemic problems. This, in turn, will stimulate greater participation in employee reporting of safety concerns.

ASAPs can be used as part of the employee reporting system for the employee groups covered by the ASAPs. However, the confidential employee reporting system required by part 5 must include all employees in the company. Companies that do not have ASAPs may consider providing employees with a hotline, a suggestion box, or information and forms for the NASA ASRS. Your online safety reporting system may provide a portal for ASRS reporting. ASRS provides certificated employees with limited immunity in the form of waivers of sanctions for reported events with certain restrictions.

It is common for organizations to treat each employee report, audit finding, or investigation in isolation. Often, system problems may not be seen if data points are examined in isolation. Thus, analysis processes should also look across individual reports and among various data sources for patterns or trends.

For example, in a recent case, examination of data from an airline's flight data monitoring program showed several instances of destabilized approaches and exceeding flap and landing gear speeds on approach. It would be easy to assume that pilot technique was the cause of the problem and then counsel or retrain the pilots.

Upon closer analysis and comparison of the events, however, the airline found that all of the instances were at one specific airport. Contacts with other airlines indicated similar experiences at the same airport. After further conversations with air traffic control (ATC), the parties concluded that the traffic handling practices were causing the problems. At that airport it was common for flights to be vectored close and high, which resulted in the approach problems. After further conversations with ATC management, the problem was resolved.

Analysis

Similar to the SRM process, SA requires analysis of the data, but this data the monitoring of current procedures and processes in place. There are many methods to make sense of the data acquired, one method to start the analysis process is as follows:

1. Establish the context: understand the safety performance objectives of the system, operations, or SMS. For system impacts, and to analyze risk controls developed under SRM, you would also need to review the system analysis conducted under SRM.

2. Identify the objective of the analysis: are you analyzing the safety performance of a system, of an operation, or the SMS itself?

3. Secure appropriate data: the data needed may be already on hand, or additional data gathering may be needed, such as conducting a special audit with focus on a specific problem.

4. Select an appropriate data analysis method: analysis need not be sophisticated to yield valuable results. For example, analysis of employee reports or qualitative analysis by SMEs may be the best method. If desired, several classification systems exist to help convert subjective, qualitative data into quantitative data for tracking and trend analysis. For routine reporting, analysis may consist of tracking such things as dispatch reliability per month, system or part failure rates, crew utilization/duty time, and events such as minor incidents, diversions, and precautionary engine shutdowns.

5. Recommendation: at this point, the person conducting the analysis may compare performance against relevant company safety objectives. Unless the decision maker is person-

ally conducting the analysis, an assessment recommendation may be made. In the case that a potential regulatory violation is discovered during analysis, the carrier may initiate a self-disclosure under voluntary reporting procedures.

6. Documentation: prepare reports and records in a format appropriate to your operation. The outputs from data analysis could be recorded in a simple recording medium such as a notebook or online system.

System Assessment

Under §5.71, collected safety performance data is analyzed and the results are used for informed decision-making. The assessment process is where these decisions are made. Personnel make decisions with assigned responsibility and authority. The SA process should consider who makes the decisions regarding whether the company's safety performance is effective and whether the company is meeting its safety objectives and expectations identified in the safety policy required by §5.21. The conclusions of the safety assessments are reported to the accountable executive, who possesses ultimate authority to act on such conclusions.

Assessments can have one of the following general outcomes:

1. Performance is acceptable and objectives are being met.
2. Performance is not acceptable, and analysis suggests that the problem lies with conformity with regulations or company policy and procedures, or necessary resources have not been provided. In the event this occurs, corrective action under §5.75 would be warranted.
3. Conformity with the risk controls and regulations appears to be satisfactory; however, desired results are not being obtained. In the event that this occurs, the SRM processes would be triggered.
4. New or uncontrolled hazards are discovered. This may be new hazards arising since the system was designed or the discovery of factors that were overlooked. In this case, as in the previous, the SRM processes must be followed.

The results of assessments can be recorded in a paper or electronic medium in a common logbook-style binder, an electronic file folder, common desktop software, or a specialized system.

Preventive/Corrective Action

The final step within system assessment is continuous improvement through preventive and corrective action(s). This process is designed to ensure that you are correcting substandard safety performance identified during the safety performance assessment in order to continuously improve safety performance. The corrective action process is triggered when conformity with risk controls has been found to be deficient. In this case, it is not necessary to conduct a new safety risk analysis. Risk has already been assessed as being unacceptable without satisfactory completion of the risk control. For example, if it has been found that an Airworthiness Directive (AD) has not been applied to a particular aircraft, the only correct action is to comply with the risk control (in this case, the AD). The risk of flying the airplane without the AD has already been assessed as unacceptable. In essence, the correction action is to ensure we are doing what we say we are doing.

This is one reason why one of the responsibilities of all employees is to follow the procedures and processes in place. When we are out of compliance we increase risk and we also make it difficult to determine if the risk controls are functioning as we planned or if those controls are ineffective.

14 CFR Part 5, Subpart D—Safety Assurance

§5.71 Safety performance monitoring and measurement.

(a) The certificate holder must develop and maintain processes and systems to acquire data with respect to its operations, products, and services to monitor the safety performance of the organization. These processes and systems must include, at a minimum, the following:

(1) Monitoring of operational processes.

(2) Monitoring of the operational environment to detect changes.

(3) Auditing of operational processes and systems.

(4) Evaluations of the SMS and operational processes and systems.

(5) Investigations of incidents and accidents.

(6) Investigations of reports regarding potential non-compliance with regulatory standards or other safety risk controls established by the certificate holder through the safety risk management process established in subpart B of this Standard.

(7) A confidential employee reporting system in which employees can report hazards, issues, concerns, occurrences, incidents, as well as propose solutions and safety improvements.

(b) The certificate holder must develop and maintain processes that analyze the data acquired through the processes and systems identified under paragraph (a) of this section and any other relevant data with respect to its operations, products, and services.

§5.73 Safety performance assessment.

(a) The certificate holder must conduct assessments of its safety performance against its safety objectives, which include reviews by the accountable executive, to:

(1) Ensure compliance with the safety risk controls established by the certificate holder.

(2) Evaluation the performance of SMS

(3) Evaluate the effectiveness of the safety risk controls established under §5.55(c) and identify any ineffective controls.

(4) Identify changes in the operational environment that may introduce new hazards.

(5) Identify new hazards.

(b) Upon completion of the assessment, if ineffective controls or new hazards are identified under paragraphs (a)(2) through (5) of this section, the certificate holder must use the safety risk management process described in subpart C of this part.

> **§5.75 Continuous improvement.**
> The certificate holder must establish and implement processes to correct safety performance deficiencies identified in the assessments conducted under §5.73.

SAFETY PROMOTION COMPONENT

Safety promotion is new to safety. Today there is an emphasis in safety training and strengthening a positive safety culture. Safety training is something not often developed under a safety program. Safety training was once more focused on facets that improved production, using the equipment properly or how to use personal protection devices. Safety culture, if discussed, traditionally was identified by "safety first" signs as well as other cosmetic signage, emblems, and company paraphernalia displayed in the buildings and on personnel's clothing.

> **Safety promotion** means a combination of training and communication of safety information to support the implementation and operation of an SMS in an organization.

Under SMS, safety promotion helps to bridge the gap between the other three components. It includes training, communication, and other actions to create a positive safety culture within all levels of the workforce. The activities within safety promotion include:

1. Providing SMS training.
2. Advocating/strengthening a positive safety culture.
3. System and safety communication and awareness.
4. Matching competency requirements to system requirements.
5. Disseminating safety lessons learned.
6. Defining how everyone has a role in promoting safety.

Your organization will need to train its employees. Employees may receive initial safety training for them to understand and perform their safety duties. They will need to know, at the end of the year evaluation, how with their safety performance be measured. Recurrent training may also be necessary to reinforce these skills.

As an organization, you will need a training plan to ensure your employees are competent. Competency is an observable, measurable set of skills, knowledge, abilities, behaviors, and other characteristics that individuals exhibit as they successfully perform work functions. Competencies are typically required at different levels of proficiency depending on the work roles or occupational function. By training your employees, you should establish competencies for all employees, commensurate with their duties relevant to the operation and performance of the SMS. Competence can be assessed at the completion of training by written, oral, or demonstration tests, and then measured periodically during the performance of that individual's work by way of periodic evaluations or supervisor/management observations. As part of SA, organizations should periodically review their training program(s) to ensure that those programs meet the objectives set out in the safety policy.

It is the responsibility of your organization to determine its own training needs based on competency requirements. Management personnel, specifically designated by the accountable executive to ensure the SMS is fully implemented, may need to be trained first and may also need specialized training to fulfill their responsibilities. Determining the organization's training needs starts with a careful review of the safety policy, processes, and objectives. Everyone working within the scope of SMS should receive training commensurate with their position in the organization.

Effective communication involves adjusting the content of the communication and manner in which the information is delivered to match the target employee's role in the organization. As an organization, when those affected are informed of the change and why the change was made, they are more likely to remember and follow the new process or procedure. The accountable executive must ensure that communication mechanisms are both available and are effectively utilized. The delivery system should be appropriate according to the size and complexity of the organization. Generally speaking, one mode of communication is not enough. To only use one mode of communication such as email results in an eventual breakdown in communication. Rather, use various planned modes to ensure the greatest level of communication across the entire system.

Safety policy and information could be provided as text, visual media such as posters or short videos, orally, or through examples. Messages should be consistent and in a form employees at each level can relate to, and be delivered using whichever media the organization has available. For example, hazard communications regarding birds for flight crew members (regarding new bird strike avoidance techniques) may be in a "Notices" section of the Flight Operations Manual (FOM), and may be reinforced by recurrent training. Hazard communications made to line maintenance technicians (regarding birds roosting and nesting in flight controls, auxiliary power unit (APU) intakes, and engine cowlings) may be conveyed by posters and changes to daily inspection procedures. Hazard communications regarding birds made to ground service personnel can be posters, videos, and demonstrations (cleaning and removing bird droppings from windshields).

You must establish and maintain SMS information, in either paper or electronic form, describing the safety-related processes and procedures and interfaces between these. You should also implement a distribution system to ensure that the documents dealing with these processes and procedures are promptly updated whenever there is a change in one or more of these processes or procedures.

Advocating for and strengthening a positive safety culture is one of the most critical components of safety promotion. Safety culture is something that cannot be mandated yet must be intentionally nurtured, promoted, and encouraged, with a mindset of continuously improving upon it. An effective way to begin this process is to measure your current safety culture. Establish a baseline regarding your safety culture climate. It doesn't matter if your safety culture is high or low, that baseline provides you with a starting point for continuous improvement. There are many available surveys to start. The answers to the questions you receive will help you to start developing a plan to improve for the next year. The goal is continuous improvement not perfection.

14 CFR Part 5, Subpart E—Safety Promotion

> **§5.91 Competencies and training.**
> The certificate holder must provide training to each individual identified in §5.23 to ensure the individuals attain and maintain the competencies necessary to perform their duties relevant to the operation and performance of the SMS.

> **§5.93 Safety communication.**
> The certificate holder must develop and maintain means for communicating safety information that, at a minimum:
> **(a)** Ensures that employees are aware of the SMS policies, processes, and tools that are relevant to their responsibilities.
> **(b)** Conveys hazard information relevant to the employee's responsibilities.
> **(c)** Explains why safety actions have been taken.
> **(d)** Explains why safety procedures are introduced or changed.

HARMONIZATION OF THE FOUR COMPONENTS

It it is critical to recognize the importance of how all four components of an SMS work in harmony with one another. Without this understanding, you will not be able to effectively develop and maintain an SMS within your organization. When operating under the principles of SMS, you cannot isolate any one component. They are all interdependent upon one another. If you have identified a hazard and established some risk controls under the SRM component, you must consider the other three components.

For example:

- How will it be written down (safety policy)?
- Who will be responsible to ensure it is implemented (safety policy)?
- How will we communicate the new safety procedure (safety promotion)?
- How will we ensure those affected are properly trained (safety promotion)?
- Will it do what we intended (safety assurance)?
- How will we measure the change's effectiveness (safety assurance)?
- When will we review the procedures to see if it worked (safety assurance)?

These are all questions that naturally get asked. As we delve into one component, we find ourselves quickly moving to the other components, as well. Simply stated, think of all four components anytime there is change as a process to manage safety and change.

With further analysis, we can see another symbiotic relationship between SRM and SA. While all four components are necessary, it is also clear that SRM and SA are key processes of the SMS. They are also highly interactive. Figure 2-7 on the next page will help you visualize these components and their interactions. The interface attribute concerns the input-output re-

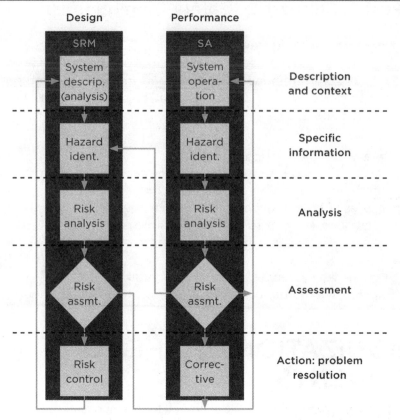

Figure 2-7 Symbiosis of SRM and SA.

lationships between the activities in the processes. This is especially important where interfaces between processes involve interactions between different departments and contractors. Assessments of these relationships should pay special attention to flow of authority, responsibility and communication, as well as procedures and documentation.

REVIEW QUESTIONS

1. What are the four components of a safety management system?

2. Explain how safety policy contributes to the entire safety management aystem.

3. Explain how safety risk management contributes to the entire safety management system.

4. Explain how safety assurance contributes to the entire safety management system.

5. Explain how safety promotion contributes to the entire safety management system. Complete "Rate Your Safety Culture Survey" online (at *https://www.tc.gc.ca/eng/civilaviation/ publications/tp13844-menu-275.htm*, see the Reader Resources page for a link). What does it say about your organization?

6. Describe in further detail how SRM and SA interface.

CHAPTER 3
SMS Costs Versus Benefits

OBJECTIVES

- Explain how effective safety management has a realistic balance between safety and production goals.
- Describe the fundamental objective in a business organization.
- Recall the current statistical value of a human life used by the Department of Transportation.
- Explain the direct costs associated with an accident or a serious incident.
- Explain the indirect costs associated with an accident or a serious incident.

KEY TERMS

- Direct cost
- Indirect cost
- Production vs. Protection
- Value of a Statistical Life (VSL)

INTRODUCTION

You need to ensure that your organization's SMS will be capable of:

- Receiving safety input from internal and external sources and integrating that information into their operational processes;
- Establishing and improving organizational safety policy to the highest level;
- Identifying, analyzing, assessing, controlling and mitigating safety hazards;
- Measuring, assuring and improving safety management at the highest level;
- Promoting an improved safety culture throughout their entire organization; and
- Realizing a return on SMS investment through improved efficiency and reduced operational risk.

This chapter will focus on realizing a positive return on your SMS investment.

IS SAFETY A CORE BUSINESS FUNCTION?

In successful aviation organizations, the management of safety is a core business function—as is financial management. We often hear aviation professionals tell us that nothing is more important than safety.

Can safety really be the number one objective? Probably not. Successful aviation organizations establish effective safety management that has a realistic balance between safety and production goals. The finite limits of personnel, time, resources, financing, and operational performance must be accepted in any industry. If properly implemented, safety management maximizes both safety and the operational effectiveness of an organization. Safety must co-exist with our production objectives. There is no aviation organization that has been created to deliver only safety.

COST VERSUS BENEFIT

A misperception has been pervasive in aviation regarding where safety fits, in terms of priority, within the organization. This misperception has evolved into a universally accepted stereotype: in aviation, safety is the first priority. While socially, ethically, and morally impeccable; the stereotype and the perspective that it conveys does not hold ground when considered from the perspective that the management of safety is an organizational process.

All aviation organizations, regardless of their nature, have a business component with production goals. An air traffic control facility may have a production goal of 100 aircraft operations per hour. An airport may have a production goal of 100 operations per hour, using parallel runways, under IFR conditions. Thus, all aviation organizations can be considered business organizations with production goals.

A simple question can shed light on the truthfulness, or lack thereof, of the safety stereotype. So, ask yourself: what is the fundamental objective of a business organization? The answer is obvious: to deliver the service for which the organization was created in the first place, to achieve production objectives, and eventually deliver dividends to stakeholders.

COST VERSUS BENEFIT CONSIDERATIONS

Operating a profitable, yet safe airline or service provider requires a constant balancing act between the need to fulfil production goals (such as departures that are on time) versus safety goals (such as taking extra time to ensure that a door is properly secured). The aviation workplace is filled with potentially unsafe conditions that will not all be eliminated; yet, operations must continue.

Some operations adopt a goal of zero accidents and state safety is their number one priority. The reality is that operators (and other commercial aviation organizations) need to generate a profit to survive. Profit or loss is the immediate indicator of the company's success in meeting its production goals. Safety is a prerequisite for a sustainable aviation business, as a company tempted to cut corners will eventually realize. For most companies, safety can best be measured by the absence of accidental losses. Companies may realize they have a safety problem following a major accident or loss, in part because it impacts the profit/loss statement. Then again, a company may operate for years with many potentially unsafe conditions without adverse consequence. Without effective safety management to identify and correct these unsafe conditions, the company may assume that it is meeting its safety objectives, as evidenced by the absence of losses. In reality, it has been lucky.

Safety and profit are not mutually exclusive. Indeed, quality organizations realize that expenditures on the correction of unsafe conditions are an investment towards long-term profitability. Losses cost money. As money is spent on risk reduction measures, costly losses are reduced (as shown in Figure 3-1). By spending more and more money on risk reduction, the gains made through reduced losses may not be in proportion to the expenditures. Companies must balance the costs of losses and expenditures on risk reduction measures. Some level of loss may be acceptable from a straight profit and loss point of view, however, few organizations can survive the economic consequences of a major accident. Hence, there is a strong economic case for an effective SMS to manage the risks.

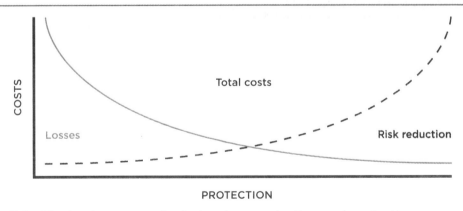

Figure 3-1   The business case for balancing production and protection.

COSTS OF ACCIDENTS

There are two basic types of costs associated with an accident or a serious incident: direct and indirect costs.

Direct Costs

These are the obvious costs that are fairly easy to determine. They mostly relate to physical damage and include rectifying, replacing or compensating for injuries, aircraft equipment and property damage. The high costs of an accident can be reduced by insurance coverage. (Some large organizations effectively self-insure by putting funds aside to cover their risks.)

Indirect Costs

While insurance may cover specified accident costs, there are many uninsured costs. An understanding of these uninsured costs (or indirect costs) is fundamental to understanding the economics of safety. Indirect costs include all those items that are not directly covered by insurance and usually total much more than the direct costs resulting from an accident. Such costs are sometimes not obvious and are often delayed. Some examples of uninsured costs that may accrue from an accident include:

- Loss of business and damage to the reputation of the organization. Many organizations will not allow their personnel to fly with an operator with a questionable safety record.
- Loss of use of equipment. This equates to lost revenue. Replacement equipment may have to be purchased or leased. Companies operating one-of-a-kind aircraft may find that their spares inventory and the people specially trained for such an aircraft become surplus.
- Loss of staff productivity. If people are injured in an accident and are unable to work, many States require that they continue to be paid. Also, these people will need to be replaced at least for the short term, incurring the costs of wages, overtime (and possibly training), as well as imposing an increased workload on the experienced workers.
- Investigation and clean-up. These are often uninsured costs. Operators may incur costs from the investigation including the costs of their staff involvement in the investigation, as well as the costs of tests and analyses, wreckage recovery, and restoring the accident site.
- Insurance deductibles. The policyholder's obligation to cover the first portion of the cost of any accident must be paid. A claim will also put a company into a higher risk category for insurance purposes and therefore may result in increased premiums. (Conversely, the implementation of a comprehensive SMS could help a company to negotiate a lower premium.)
- Legal action and damage claims. Legal costs can accrue rapidly. While it is possible to insure for public liability and damages, it is virtually impossible to cover the cost of time lost handling legal action and damage claims.
- Fines and citations. Government authorities may impose fines and citations, including possibly shutting down unsafe operations.

COSTS OF INCIDENTS

Serious aviation incidents, which result in minor damage or injuries, can also incur many of these indirect or uninsured costs. Typical cost factors arising from such incidents can include:

- Flight delays and cancellations;
- Alternate passenger transportation, accommodation, complaints, etc.;
- Crew change and positioning;
- Loss of revenue and reputation;
- Aircraft recovery, repair and test flight; and
- Incident investigation.

COSTS OF SAFETY

The costs of safety are even more difficult to quantify than the full costs of accidents. This is partly because of the difficulty in assessing the value of accidents that have been prevented. Nevertheless, some operators have attempted to quantify the costs and benefits of introducing an SMS. They have found the cost savings to be substantial. Performing a cost-benefit analysis is complicated, however, it is an exercise that should be undertaken, as senior management is not inclined to spend money if there is no quantifiable benefit. One way of addressing this issue is to separate the costs of managing safety from the costs of correcting safety deficiencies, by charging the safety management costs to the safety department, and the safety deficiency costs to the line management most responsible. This exercise requires senior management's involvement in considering the costs and benefits of managing safety.

In successful aviation organizations, safety management is a core business function - as is financial management. Effective safety management requires a realistic balance between safety and production goals. Thus, a coordinated approach in which the organization's goals and resources are analyzed helps to ensure that decisions concerning safety are realistic and complementary to the operational needs of the organization. The finite limits of financing and operational performance must be accepted in any industry. Defining acceptable and unacceptable risks is therefore important for cost-effective safety management. If properly implemented, safety management measures not only increase safety, but also improve the operational effectiveness of an organization.

WHAT IS THE VALUE OF A HUMAN LIFE?

The United States government has conducted several studies on the treatment of the economic **value of a statistical life (VSL)**. The FAA is organized under the Department of Transportation (DOT), so we will concentrate our emphasis on the DOT's value of a statistical life.

The DOT recognized VSL has a major effect on policy for the FAA. Federal statutes established by the United States Congress require the FAA to promote safety and air commerce. Ultimately this requires a cost/benefit analysis prior to making a new regulation. In contrast, the National Transportation Safety Board (NTSB) does not conduct cost/benefit ratio analysis prior to issuing recommendations to the FAA.

The United States Department of Transportation (DOT) guidance on valuing reduction of fatalities and injuries by regulations or investments has been published periodically since 1993. The most recent empirical study published in 2012 indicate an average value of a statistical life (VSL) of $9.1 million in U.S. dollars for analyses. Although the average value of a human life is $9.1 million, the DOT has established a variability of +/- $3.8 million. This ultimately establishes a range between $5.2 million up to $12.9 million.

The new 14 CFR Part 5 used the value of a statistical life at $8.9 million based on 2010 dollars.

VALUE OF PREVENTING INJURIES

An accident that results in a loss in quality of life, including both pain and suffering and reduced income, should also be estimated. The dollar value for being injured will be less than the rate for the loss of a life. The fractions shown in the following table should be multiplied by the current VSL to obtain the values of preventing injuries of the types affected by the government action being analyzed.

For example, if the analyst were seeking to estimate the value of a serious injury (AIS 3), he or she would multiply the Fraction of VSL for a serious injury (0.105) by the VSL ($9.1 million) to calculate the value of the serious injury ($955,000). Values for injuries in the future would be calculated by multiplying these Fractions of VSL by the future values of VSL.

AIS Level	Severity	Fraction of VSL
AIS 1	Minor	0.003
AIS 2	Moderate	0.047
AIS 3	Serious	0.105

Table 3-1 Relative disutility factors by accident injury severity level (AIS).

RECOGNIZING UNCERTAINTY

Multiple studies have been conducted to determine the value of a human life. Some studies suggest a reasonable range of values for VSL between $4 million and $12.9 million. Additionally, different organizations within the U.S. government use different values.

The relative costs and benefits of different provisions of a rule can vary greatly, therefore it is important to disaggregate the provisions of a rule, displaying the expected costs and benefits of each provision, together with estimates of costs and benefits of reasonable alternatives to each provision.

PRODUCTION VERSUS PROTECTION

In most organizations safety is the number one priority. That is what the public expects, and that is what most (if not all) organizations would like you to think, especially in aviation. What aviation organization is not going to tell you that safety is their number one priority? If there is one, they are probably not going to be in business very long. Nobody is going to want to do business with an organization that admits that safety is not its number one priority, especially if the customer is going to be a flying passenger!

But, saying that safety is your top priority and actually making it your top priority are two, totally separate items. Is safety really the number one priority for an organization? We would like to think that this is true, but in reality it probably is not going to work out that way in the long run.

So, why not? To really answer this question we have to look at what the fundamental objective of any company or organization actually is. The fundamental objective of any organization is to meet their production goals or objectives. If they are a for profit corporation this could be taken one step further to say that that objective is to make a profit.

In either of those cases, is safety a competing priority? This can depend on the company. A quick definition of safety is: "the absence of risk." If an aviation organization is going to be completely absent from risk, they are going to have stop flying their aircraft, because flying definitely involves risk. They are probably going to have to go one step further and park the airplanes in a remote area of the airport with fence around them to prevent people from gaining access to them and having the possibility of injuring themselves. Okay, the last part may be a bit much, but you get the idea.

Of course, the problem with this scenario is that it is unrealistic. If we park all our airplanes, we are not going to meet any of our goals or objectives, including making money. So, what can we do about this? One thing we are going to have to do is accept that there will be some risk in our operation. We are going to have to find a way to manage that risk, thereby letting us work on meeting our production objectives.

Following this logic it seems that safety may not be able to be our top priority. But, can it be complementary? There are plenty of airlines flying today, and the general public does not think twice about flying from point A to point B on them. But, airlines do not use safety in their marketing efforts. Why is it that passengers still fly on these airlines? It is because safety is assumed. Passengers expect that they will be safe on any given airline that they fly on. This is probably a realistic expectation, at least in areas of the world that have very robust oversight into airline safety.

In a company where safety is taken seriously it is treated as a core business function, much like accounting or marketing. Proper safety management will be able to properly manage the risk associated with operating aircraft or whatever the organization is doing. We need to find the "sweet spot" where the right balance is achieved between production and protection (see Figure 3-2.)

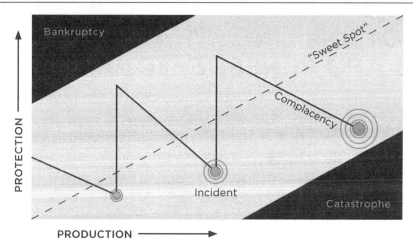

Figure 3-2   Achieving balance between production and protection.

If a corporation is to achieve the organizational objectives, but also be a safe organization, it will have to commit some of its resources into safety management. Just because an organization commits financial resources to safety, does not mean that the desired level of safety is achieved. The members of the organization from top to bottom have to believe in the safety objective. How is this achieved? Through a strong safety culture, that's how.

BENEFITS OF MANAGING SAFETY

Clearly, the ultimate goal of SMS is increased safety—in particular, fewer accidents and injuries. Moreover, increasing a system's level of safety leads to reduced material losses and enhances productivity. This makes the case that safety is good for business. Some further benefits include:

- Reduction of the direct and indirect costs of accidents. Fines, repair costs, damage claims, and increased insurance premiums are a few of the potential economic consequences of an aviation mishap.
- Improved employee morale and productivity. Promoting communication between management and the rest of the organization prevents disenfranchisement and lifts morale.
- Establishing a marketable safety record. A record of consistently safe operations can be used to attract new business and investment.
- Logical prioritization of safety needs. SMS emphasizes risk mitigation actions that provide the biggest impact on both safety and the bottom line.
- Compliance with legal responsibilities for safety. Airport certification requirements mandate a number of safety processes and standards that can be included in an organization's SMS.
- More efficient maintenance scheduling and resource utilization. Effective hazard reporting in SMS allows proactive scheduling of maintenance tasks when resources are available, increasing the likelihood that maintenance is performed on time and more efficiently.
- Avoiding incident investigation costs and operational disruptions. Improved communication and risk mitigation will prevent many accidents from ever occurring.
- Continuous improvement of operational processes. SMS allows for lessons learned to be incorporated into the system and lead to superior operations.

THE BOTTOM LINE: WHAT ARE THE COSTS AND BENEFITS OF PART 5?

With the implementation of Part 5, each air carrier is required to develop an SMS that includes the four SMS components: safety policy, safety risk management, safety assurance, and safety promotion. To support each component, the FAA projects that the compliance cost of this rule will come from the initial development and documentation of the carrier's SMS, implementation and continuous operating costs to include the modification or purchasing of new equipment/software, additional staff and promotional materials, and training. Costs increase with the size of the carrier and the type of operations that they provide. Medium and large operators have existing quality management systems that will lower their estimated compliance costs.

The FAA estimates that the benefits from Part 5 could be between $205.0 and $472.3 million over 10 years ($104.9 to $241.9 million present value at 7 percent discount rate). The costs of the rule's provisions (excluding any mitigation costs, which have not been estimated) are estimated to be $224.3 million ($135.1 million present value at 7 percent discount rate) over 10 years.

REVIEW QUESTIONS

1. True or false: Not a single aviation organization has been created to deliver only safety.

2. How important is customer's perception of safety to an airline?

3. What could be the result of committing too many resources to safety?

4. Which of the following statements concerning cost/benefit ratio analysis is true?
   a. The "benefit" portion of cost/benefit ratio analysis includes loss of revenue, installation costs, and maintenance costs.
   b. The statistical value of a human life is determined by a congressional committee. This exact same statistical value of a human life is then used for all government agencies. (As an example: The DOT and the EPA use the same value.
   c. The FAA is required to conduct a cost/benefit analysis prior to making any new regulations.

5. According to the DOT, what is the current statistical value of a human life?

6. True or false: All United States Government agencies forced to use the same statistical value of a human life. For example, the DOT and the EPA both use $8.0 million as the value of a human life.

7. True or false: Safety is widely used as an advertising and marketing tool with almost every air carrier.

8. True or false: The FAA is allowed (because of laws passed by congress) to routinely establish new regulations that would result in the bankruptcy of most aviation companies.

9. Which of the following are the charters given to the FAA by the U.S. Congress:
   a. Promote a zero mishap rate.
   b. Implement a "Zero Delay Program" within the Air Traffic Control System.
   c. Promote air commerce and air safety.
   d. Provide government oversight of the "Lost Baggage Program."

CHAPTER 4
Safety Management Systems Versus Safety Programs

OBJECTIVES

- To analyze the traits of a safety management system and compare it to a safety program.
- To evaluate various scenarios and determine whether or not they reflect SMS.

KEY TERMS

- Key Safety Management Personnel
- Processes
- System
- Unintended Consequences

INTRODUCTION

One of the best ways to identify if an organization or an individual understands the concepts of an organization's SMS is to ask one simple question: "Tell me the difference between a safety program and a safety management system?" The response will vary depending on the complexity of an organization's current safety program. Normally individuals describe a safety management system based on the positive aspects of their current safety program and culture.

They don't know, what they don't know. A manager may say, we have a safety reporting system, we have a safety office, and a common response from management, "If people have a safety concern they know my office is always open." The focus is on the appearance; the clean floors, the "safety first" signs, the metal coins that are handed out to celebrate individual success. While all these appearances on the surface are commendable, they are not what differentiate a safety program from a safety management system.

The FAA defines SMS as "the formal, top-down business approach to managing safety risk, which includes a systemic approach to managing safety, including the necessary organizational structures, accountabilities, policies and procedures." At first glance, this is a daunting description, but as we break it down we can learn much from the definition.

BREAKING DOWN THE DEFINITION OF SMS

"The formal, top-down business approach..."

SMS is all about safety decision-making throughout the organization. It is formal in that it requires conventional requirements for organizational behavior and procedures. It is top down because of the fact-based decision-making that occurs through a system that measures safety objectively. Decision makers become your **key safety management personnel** within your organization. Risk accountability moves from the safety department to the chief executive officers, vice presidents, and directors who are responsible for various departments and **processes** within an organization.

SMS takes a business approach in that those decision makers, who are also responsible for accepting and mitigating risk, understand that safety is good for business. They also understand some risk is necessary because if risk is not accepted the organizations mission cannot be met. For example, it would be less risk if all airline flights had two pilots that met captain requirements or if all 14 CFR Part 91 corporate flight departments had two pilot only operations. In flight training, it would be safer to never let a student pilot fly solo, or allow a student to fly at night. While these situations lessen risk, clearly you need them to happen to fulfill the mission. An airline can't afford two captains on all routes; some corporate pilots can safely fly without a co-pilot and the flight training school must ensure each student is career-ready at graduation.

"Managing safety risk, which includes a systemic approach to managing safety…"

Key safety management personnel, who are your organizational leaders, cannot use "mission" to allow for acceptance of unnecessary or unneeded risk. In fact, they are responsible and held accountable to ensure that risk is reduced to the lowest practical level.

What does this mean? It means that as we recognize risk within our organization, we must practically balance the need for meeting our mission as an organization with purposefully identifying and managing risk, through a systemic approach. In later chapters, we will explore the concept of who can accept different levels of risk within an organization. For now, know that normally the higher the risk, the more senior the organizational risk acceptance must be deferred to.

To really understand a systemic approach to managing safety and really compare a safety management system to a safety program, you must first understand the system. If you do not fully grasp what a system is, you will find it difficult to grasp the full significance of a safety management system. "System" is a term that is used so often in the FAA and ICAO publications yet it is commonly overlooked and the significance is missed by most.

According the Safety Management System Voluntary Guide, a **system**, means "a group of interacting, interrelated, or interdependent elements forming a complete whole." Explained in plain language, the system is how all the elements or parts work together to create the whole. That whole does not function or exist without the interaction of all the parts. Each part, no matter how flawed still makes up the whole. If you change even the most flawed part of the system, it will affect the remaining parts. The parts that work well will only work well, in part, because they are working in conjunction with those other flawed parts. Change any part and it will affect the other parts and hence the system.

For example, say you are building a new house, and you have hired a professional architect to design the best house possible. The house is magnificent; each room has a specific size and purpose. Each part fits well with the intent you have for its use. When the designs are complete and you think you have a finished product, you look at your spouse and they say, "the corner room needs to be twice as large." You cringe. In your mind you think, "will the benefit outweigh the cost?" You know you want it done right, so you say, "fine, just make the improvement, it will be better than what it was and it will hopefully make my spouse happy." Fortunately, before continuing, you consult your architect who looks at the house as an entire system, a whole. The architect doesn't see parts as individual pieces, but sees an overall house that has connecting parts. He indicates he will not just expand the room, but will need to consider how that change will affect the entire structure of the house as well as the connecting rooms. It makes sense. If I change this one room it will have a significant effect on the adjacent rooms, and depending on how it was changed, it may affect other rooms, changing their purpose too.

The house represents the system; each room represents the parts that make up that house or system. The rooms could be different departments, the furniture could be the policies, and the doors and stairs represent the processes in place. If any part changes, it has the possibility to affect the system. Before approaching any change, we need to always think of the entire system.

Taking this further, one who understands the system would recognize the danger of changing just because "it will be better than what is was." Be careful what you ask for. A change to one part, while it may improve that "room," has the potential to affect the other rooms and subsequently the entire system for better or worse.

Let's imagine your flight school has an abnormal amount of loss of directional control (LODC) incidents that occur during high crosswind landings. The FAA is putting pressure on your organization to quickly make a change, to fix this problem. Without considering the system, your chief instructor decides to appease the FAA and reduces the maximum cross wind component from 20 knots to 10 knots. The FAA applauds your chief instructor's quick and decisive solution. It isn't until six months later when your organization begins to experience the negative and unintended consequences associated with a decision that was made with no regard to the entire system. Later chapters will delve more into unintended consequences, how to identify them, how to minimize them, and how to correct for those you did not anticipate.

"Including the necessary organizational structures, accountabilities, policies and procedures."

Many have heard the term "corporate knowledge." This term relates to what an organization knows collectively but is not documented or in any way recorded. For example, "everyone knows if an aircraft is damaged, you need to contact the director of maintenance." Phrases such as, "that's how we have always done it" or "here's what we did last time this happened" are used. SMS requires that we have organizational structures and processes in place that are documented and clearly give direction and communication to the entire organization. Processes are in place to identity hazards, to analyze and assess risk, and to mitigate risk.

Those organizational structures address all four components of a safety management system. Accountabilities, policies, and procedures must also be documented, communicated, and trained. Organizational structures must be established that identify who is accountable for safety, who can accept risk, and who can delegate responsibility within the organization. Policies and procedures must be established that clearly lay out the safety objectives of the organization as well lay the foundation for what is expected throughout the organization, from the leaders to the part-time employees.

CHARACTERISTICS OF AN SMS VERSUS A TRADITIONAL SAFETY PROGRAM

There are several characteristics that separate an SMS from a traditional safety program (see Table 4-1).

Safety Management System	Traditional Safety Program
Management's commitment to safety is by active involvement in giving direction to reduce risk and accept risk.	Management's commitment to safety is done through verbal encouragement but rarely with personal action or involvement in accepting risk.
Non-punitive safety reporting is available for all personnel to voice their safety concerns.	Safety reporting is often available, but has been or has the potential to be punitive to the individual who reports.
Personnel feel comfortable to self-report without fear of reprisal.	Safety reports are only done if there is assurance that it is anonymous and untraceable to reporter.
Trend analysis is used to identify hazards and make fact-based decisions.	Trend analysis results in admonishment of employee performance and is not part of the decision-making process for change.
Safety reporting normally expands to multiple vendors or sub-contractors to promote identification of hazards for the organization as a whole.	Safety reporting data is keep secure and not available for most personnel.
Recognizes the importance of the technical experts, who must follow procedures and policies every day.	Often will publish policies with little input from personnel.
Processes to proactively identify of hazards are documented and tracked.	Reactively identifies hazards after an occurrence or incident with little or no documentation.
Processes are documented to predictively anticipate hazards and associated risks.	Normally not accomplished in a safety program. If data is available, it is rarely reviewed unless there is cause, such as an accident, incident, or report.
Processes are in place to manage risk as hazards are identified.	Hazards are addressed when they are identified and then considered corrected.
Processes are in place to ensure actions taken to reduce risk are performed as designed.	Normally not accomplished in a safety program.
Processes are in place to correct actions taken to reduce risk if they are not performing as designed.	Normally not accomplished in a safety program.
Safety management is considered essential for the business to perform satisfactorily and resources are allocated as appropriate.	Safety is included, but little or no resources are allocated.
Personnel are aware of their responsibility in regard to managing safety.	Safety is the responsibility of the safety department not the operations personnel.
Decision makers are evaluated on production and protection performance measures.	Safety department personnel are evaluated only on protection. Decision-makers are evaluated only on production.
A structured means of safety risk management decision-making is employed, which takes into account the entire system and draws from subject matter experts in departments that are affected.	Key leadership, often with extensive experience, makes the decision.
Has a means of demonstrating safety management capability before system failures occur.	If there is a safety issue, personnel know through "corporate knowledge" to contact the safety department.
Exhibits increased confidence in risk controls though structured safety assurance processes.	Normally not accomplished in a safety program.
Has an effective interface for knowledge sharing between the regulator and certificate holder/organization.	Safety information details are not shared in fear of regulator penalties or violation.

Table 4-1 Differences between an SMS and a traditional safety program. *(Continued on next page)*

Safety Management System	Traditional Safety Program
Encompasses a safety promotion framework to support a sound safety culture that is measured, tracked and seeks to improve the safety culture.	Normally safety culture is unknown and not measured. Most safety programs would say they have a "strong" safety culture but would be unable to show evidence of that culture.
When accident, incident, or occurrence happens it is recognized that most likely organizational factors caused the event.	When accident, incident, or occurrence happens it is recognized that most likely, the persons involved caused the event.
The system is defined and taken into consideration before changes are made.	The system is rarely defined and change occurs without consideration to the entire system.
Contains all four components of SMS—safety policy, safety risk management, safety assurance, and safety promotion.	Contains some aspects of safety risk management and safety policy, but rarely addresses safety assurance and safety promotion.

Table 4-1 Differences between an SMS and a traditional safety program. *(Continued from previous page.)*

SUMMARY

SMS requires the organization to examine its operations and the decisions around those operations. SMS allows an organization to adapt to change, increasing complexity, and limited resources. SMS also promotes the continuous improvement of safety through specific methods to predict hazards from employee reports and data collection. Organizations then use this information to analyze, assess, and control risk. Part of the process includes the monitoring of controls and of the system itself for effectiveness. SMS helps organizations comply with existing regulations while predicting the need for future action by sharing knowledge and information. Finally, SMS includes requirements that enhance the safety attitudes of an organization by changing the safety culture of leadership, management, and employees. All these changes are designed to move the organization from reactive thinking to predictive thinking.

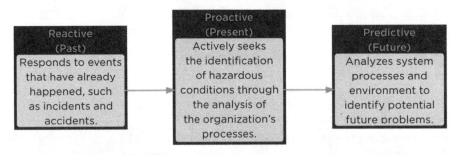

Figure 4-1 Predictive thinking.

SMS has generated wide support in the aviation community as an effective approach that can deliver real safety and financial benefits. SMSs integrate modern safety concepts into repeatable, proactive processes in a single system, emphasizing safety management as a fundamental business process to be considered in the same manner as other aspects of business management. The structure of SMS provides organizations greater insight into their operational environment, generating process efficiencies and cost avoidance. Some participants have found that benefits begin to materialize even in the early reactive stages of implementation. This continues as organizations evolve from reactive to proactive and predictive phases.

SCENARIO FOR GROUP DISCUSSION: DEMONSTRATING THE NEED FOR AN SMS

Read the following scenario from the FAA. Assume this organization has a "strong" traditional safety program. Discuss the questions listed after the scenario.

A well-designed aircraft with a history of reliable service is being prepared for a charter flight. Employees tow the aircraft from the hangar to the terminal. One employee sees wetness on the right tire as he unhooks the tow bar. However, he does not give it attention, as he is very busy and has three other aircraft to move in the next 15 minutes.

At the same time, a safety inspector is walking through the hangar when she encounters a hydraulic oil spill on the hangar floor. She notifies a janitor to clean up the slip hazard as she leaves. While cleaning the spill, the janitor wonders aloud where the spill came from. Afterwards, both the inspector and the janitor continue with their respective jobs.

Meanwhile, the chief pilot assigns the charter flight to a new pilot with the company. While new to the company, the pilot is well-trained and prepared for the flight. He is also eager to do a good job and to impress the chief pilot. The chief tells him that the passengers and the aircraft are waiting at the terminal, and the new pilot has to get over there right away to keep the clients happy and on schedule.

The flight requires a little more fuel, so a fuel truck is called. While the aircraft is being filled, the fueler notices a small puddle of reddish fluid under the right main landing gear. He sees the pilot walking out to the aircraft, but before he can say anything, his supervisor calls and tells him to get right over to another aircraft. Recently, the fueler was criticized by his supervisor for taking too long to finish his work, so he quickly jumps in his truck and drives off to the next job without saying anything to the pilot.

The pilot, wanting to make a good impression on his passengers and the chief pilot, personally escorts them to the aircraft and begins his preparation for the flight. One passenger asks him a brief question as he is on the right side of the aircraft. In a moment of distraction, he does not bend down to inspect the right main landing gear.

During taxi, the pilot feels the aircraft is taking the bumps a little hard, but continues to the runway for takeoff. Meanwhile, up in the tower, an air traffic controller, who happens to like this particular model of aircraft, picks up her binoculars to take a look at the taxiing aircraft. She notices a wet spot on the right main tire and radios the pilot. The pilot tells the controller that he probably ran over a puddle and asks for his clearance.

At the destination airport, the pilot executes a perfect landing and applies the brakes. The leaking hydraulic fluid heats up and ignites. The right main landing gear is engulfed in flames. The controller notifies the pilot and then calls the crash fire rescue squad. The pilot calmly and proficiently manages the situation, successfully evacuating everyone from the aircraft without injury. The pilot and passengers watch from a safe distance while a perfectly good aircraft burns to the ground. "How could this have happened?" wonders the pilot.

Soon afterwards, the pilot is fired for failure to perform an adequate preflight inspection. Six months later, an aircraft is being towed out of a hanger. One of the employees sees wetness on the left main landing gear tire as he unhooks the tow bar...

1. How many opportunities were there to prevent this accident?
2. Why did this happen?
3. Identify at least three items listed in the SMS column in Table 4-1 beginning on page 53 that were not evident in this scenario.
4. Identify at least three items listed in the traditional safety program column in Table 4-1 beginning on page 53 that were evident in this scenario.

REVIEW QUESTIONS

1. How does your company or organization determine who accepts risk?

2. What are some risks your organization must take in order to accomplish its mission?

3. In regard to risks your company must to take, what visible ways to they try to mitigate the risk to the lowest practical level?

4. Can you think of a change that was made to your organization in which the "system" was not taken into account? How negative and or unintentional consequences resulted?

5. What are some examples of "corporate knowledge" that you have experienced in your workplace?

6. Describe some organizational structures or processes that are in place where you work that help the organization manage safety.

Listed below are various examples of a how a typical safety program would respond to a situation. Describe how an organization with an SMS would respond.

7. The safety department has been conducting no-notice observations of ramp operations throughout the entire day. It has been observed that line vehicles are often traveling at speeds above the policy speed limits. The V.P. of aviation safety sends an email to all line personnel stating, "It has been observed that many employees are operating vehicles in violation of 3.4.5 of the Company Safety Operation Procedures. Please adhere to these requirements."

8. The flight operations department has experienced multiple flights where the final checklist was not completed. Improper checklist procedures have the potential to cause injury to pilots and line personnel. Therefore, the flight operations department has established a no tolerance policy regarding checklist usage. The next violation results in an experienced pilot being placed on unpaid suspension for five days.

CHAPTER 5
Scalability of SMS

OBJECTIVES

- Analyze ways a safety management system can be scalable to the scope and size of an organization.
- Give clear examples of how SMS principles can be used on a large and small scale.

KEY TERMS

- Aviation Service Provider
- FAA Gap Analysis Tool
- Flight Operations Quality Assurance (FOQA)
- Flight Standards District Office (FSDO)
- Internal Evaluation Program (IEP)
- Line Operations Safety Audit (LOSA)
- SMS Program Office (SMSPO)

INTRODUCTION

A small Part 135 operator made its initial meeting with the FAA to begin the process of being recognized by the FAA for operating with a safety management system. The FAA from the national level known as the **SMS Program Office (SMSPO)** joined with the local FAA management office known as the **Flight Standards District Office (FSDO)** came together to lay out all the details, expectations, and requirements needed to meet SMS requirements for the SMS Voluntary Program. After all morning reviewing these requirements, the operator raised his hand to ask one question, "Is SMS required for Part 135 operators?" The FAA explained that it was voluntary and they all broke for lunch. The meeting reconvened after lunch and, unsurprisingly, the Part 135 operator was absent.

The fear of SMS not being scalable, of it becoming an unmanageable behemoth of meetings and paperwork, sent the Part 135 operator home with no desire to initiate a safety management system. Unfortunately, the FAA did not get the chance to explain that SMS was meant to be scalable, to be able to conform to the size and complexity of an organization.

The FAA does recognize the perceived impact that this rule may have on small businesses. As of January 6, 2012, there were 90 Part 121 certificate holders. The size, scope, and complexity of the operations of each of these certificate holders varies greatly. For example, one-third of the Part 121 certificate holders have 10 or fewer airplanes, while 10% have more than 270 airplanes. Given the variance in these types of operations, the FAA designed these SMS requirements to be applicable to air carriers of various sizes, scopes, and complexities, as well as adaptable to fit the different types of organizations in the air transportation system and operations within an individual air carrier. The FAA does not anticipate, nor expect that small air carriers or other service providers would require an SMS as complex as one for large air carriers. To further clarify this issue, the FAA has revised 14 CFR §5.3 in the final rule to state that the SMS must be appropriate to the size, scope, and complexity of the certificate holder's operations. As such, it is scalable to the size of a small entity.

The FAA has also revised the guidance material that was published for comment with the Notice for Proposed Rule Making (NPRM). The revised guidance material provides a variety of examples of how to implement the SMS processes and procedures that an air carrier may develop based on the size, scope, and complexity of its operation. The examples outlined in the guidance material were not intended to limit an air carrier to only these methods of compliance.

DEFINING SCALABILITY

14 CFR Part 5 and the SMSVP specifies a basic set of processes integral to an effective SMS but does not specify particular methods for implementing these processes. In other words, the regulation defines what must be accomplished, not how it must be accomplished. FAA publications provide additional guidance on how the SMS may be developed to achieve the safety performance objectives outlined by your organization. Within this chapter we will demonstrate how there is no one-size-fits-all method for complying with the requirements of Part 5 or the SMSVP. This design is intentional in that the FAA expects each air carrier, corporate depart-

ment, flight operation, maintenance repair station, or other any **aviation service provider** to develop an SMS that works for its unique operation. Thus, this chapter provides guidance regarding designing and implementing acceptable methods of compliance regardless of the size of your organization. These methods, however, are not the only means of compliance.

Scalability applies to businesses that are considering applying for, have applied for, or hold a Part 121 certificate. §5.1 requires a Part 121 certificate holder to have an SMS that meets the requirements of Part 5 and that is acceptable to the Administrator of the Federal Aviation Administration (FAA). It also applies to other Aviation Service Providers who meet ICAO requirements under the SMSVP guide.

The difference between a large, medium, and small organization's SMS is primarily:

- Complexity of the operations to be covered;
- Volume of data available;
- The size of the employee workforce; and
- The resources needed to manage the organization.

The SMS requirements are the same regardless of the size of your organization. Part 5 and SMSVP guidance, however, allow organizations of different sizes to meet those requirements in different ways. The SMS functions do not need to be extensive or complex to be effective. All businesses, regardless of size, may use existing systems, programs and resources to document and track safety issues to resolution.

One of the most effective tools to measure what you already have in your organization compared to what you must still accomplish to meet SMS standards is the **FAA gap analysis tool** (see Figure 5-1 on the next page). This optional tool based on 14 Code of Federal Regulations (CFR) Part 5 and is not regulatory. It is provided by the AFS-900, SMS Program Office to aid aviation service providers in the evaluation of their existing systems and programs as compared to SMS requirements in 14 CFR Part 5. If there are discrepancies between this tool and Part 5, Part 5 shall prevail; if your organization is under the SMSVP, then the SMSVP will prevail because you are not under 14 CFR Part 5. Even though this is the case, it is advised that any discrepancies be discussed by your FSDO who will be the final authority on what is expected to meet SMSVP standards.

The gap analysis tool will help you organize your existing processes as well as develop your implementation plan in order to be in full compliance with Part 5 or the SMSVP guide. Let's further explore the factors that identify the differences between small, medium, and large operations. Operations could be air carrier, corporate flight operations, a small flight school, or an unmanned aircraft systems organization.

Size of Operation

Regardless of the size of an operation, SMS principles can be applied. If you are small and have not established any safety processes or procedures then keep it small. Document what you do have, and grow as you see the need and identify where the process is practical and beneficial for managing safety within the organization. For example, non-punitive safety reporting is a requirement under SMS. Before executing, determine the purpose of the SMS requirement. If you understand the purpose, it will prevent the likelihood of taking the requirement beyond its pur-

NPRM Part 5 - Gap Analysis Tool					Date:	

Note: This optional tool is based on 14 CFR part 5 NPRM and **is not regulatory**. It is provided by the AFS-900, SMS Program Office to aid aviation service providers in the evaluation of their existing systems and programs as compared to SMS requirements in 14 CFR part 5 NPRM. **This is a transition tool for service providers to use until 14 CFR part 5 is published.** If there are discrepancies between this tool and part 5, part 5 shall prevail.
Column headings (B thru J), the cells 'gray' area, and columns/rows may be modified, added or deleted to reflect a service provider's organizational structure.

Please make no text changes to the questions, however notes and/or comments may be added or other modifications made as necessary.

Participant:					Location:	
Regulatory Requirement Questions	Company's Documentation Source	Quality Control	Training	Maintenace Operations	Additional Comments and/or Action needed to reach level 4	
Regulatory Requirements						
Does your safety management system have a safety policy that includes at least the following minimum requirements:						
• Your organization's safety objectives,						
• A commitment to fulfill your organization's safety objectives;						
• A clear statement to commit the necessary resources for implementing your safety management system;						
• A safety reporting policy that defines requirements for your employee's to report safety hazards or issues;						
• A policy that defines unacceptable behavior and conditions for disciplinary action; and						
• An emergency response plan that provides for the safe transition from normal to emergency operations in accordance with the requirements of 14 CFR part 5.27, Coordination of Emergency Response Planning?						

Figure 5-1 Excerpt of an FAA gap analysis tool. *(FAA)*

pose or intent. Non-punitive safety reporting's purpose is to provide a means for the technical experts within your organization to report any potential safety hazards or hazard consequences experienced. It needs to be non-punitive because if employees fear punishment for self-disclosure, they will not report. If they do not report, the organization will not have valuable data that it can use to track organization trends and make fact-based decisions about its program.

Based on the size, safety reporting could be done by simply communicating and documenting the process for employees to drop a hand written letter or small form in a safety box mounted on the wall. A person(s) who is preferably not the direct supervisor of the employee could collect the data. This person can categorize and trend the safety reports, follow up as appropriate, and communicate to the employees of the actions taken. A large organization, such as a Part 121 airline, will have already developed safety reporting over decades and most likely are under an **Aviation Safety Action Program (ASAP)**. Smaller aviation service providers have also developed letters of agreement with the FAA to initiate ASAP programs, but it is predominately utilized in Part 121 operators.

An established ASAP provides a partnership between the organization and the FAA to identify hazards through safety reports, mitigate the risk while giving reasonable protection (non-punitive) to the employees who self-disclose. These large organizations will often outsource to organization who specialized in online safety reporting systems that have many layers of reporting points specific to a departments or job function.

Regardless of the size of your organization, you can customize your approach to meet what must be accomplished under a safety management system.

Complexity of Operations

The less complex an operation, the simpler the SMS processes will be. As discussed in Chapter 3, properly understanding your system will help ensure you properly address hazards within your organization. Normally in small organization, a majority of the employees understand their system. In a small airport, you have two to three full-time employees and a hand full of part-timers. Everyone is keenly aware of their environment and the system in which they operate. As complexity grows, technical experts become more specialized and it is not common that the operations personnel have little understanding of how maintenance, line operations, or other departments interact and relate to each other's area of expertise. This complexity adds a level complexity for large operators. Turnover rates are higher and people are more likely to operate independent of other departments on a micro and sometimes a macro level. Greater detail in processes must be documented and communicated.

Volume of Data Available

SMS requires fact-based decisions and data helps decision makers make informed decisions. How much data your organization is able to collect and process will affect how you approach data volume. Unmanned aircraft have a significant amount of raw data that can be accessed on the aircraft. Telemetry and the datastream from the communications downlink can be a smorgasbord of information for operators and organizations. Flight schools with glass cockpits have the ability to collect terabytes of information on ever student's flight.

Size of Employee Workforce

Larger size organizations have an advantage at initial implementation of SMS because they can absorb the initial efforts to initiate the SMS process. Airlines have already established ASAP and **Flight Operations Quality Assurance (FOQA)** programs to identify hazards and reduce risk. This means the transition from strong safety programs to a safety management system can be must less painful. Small organizations, while less complete, suffer from a simple lack of resources. Individuals who have little training in safety are thrust into an environment where the technical language becomes difficult to translate.

Resources Needed to Manage the Organization

Safety resources require significant buy-in at the early stages of SMS. Most likely resources may be re-allocated, but new resources are rarely provided without significant justification and measurable documentation of what benefit will be received by having additional safety personnel. Often safety personnel are considered a drain on the system and more safety reduced profit margins. To obtain resources, leadership must be convinced that what is being done is making the business better while improving the overall operation. Small organizations can go years before additional resources will even be allocated to safety type personnel.

SCENARIOS FOR GROUP DISCUSSION: SCALABILITY

Understand how different-sized carriers and other aviation or non-aviation service providers can meet the pertinent SMS requirements is critical for implementation. For discussion purposes, the following scenarios are broken down into small, medium, and large carriers. Regardless of what type of service provider your organization is, these scenarios will help to compare and contrast various organizations size, complexity, and other factors affecting scalability.

- **Small** carriers are generally defined as those carriers operating fewer than 10 airplanes.
- **Medium** carriers are those with fewer than 48 airplanes.
- **Large** carriers are those with more than 48 airplanes.

Safety Policy Scalability

Safety policies are expected to vary between small, medium, and large organizations, additionally, the levels of management involved in preparation and implementation of the policy may vary.

- **Small.** The owner or most senior manager (the accountable executive) may personally perform this process. The policy can be a simple, often single-page, written document, signed by the accountable executive. Small organizations typically operate in smaller networks of employees, so the policy may be posted in company work areas or included in company briefings or in training.
- **Medium.** The accountable executive, with the involvement of other senior managers, is likely to develop, publish, and communicate the safety policy. The policy may be disseminated via company newsletters, company websites, employee briefings, or existing indoctrination and recurrent training.
- **Large.** A large organization may require the accountable executive or other senior managers and technical staff to perform this process. While the regulations only require the accountable executive to sign the safety policy, members of senior management may also sign the safety policy. Large carriers may disseminate their policy using a variety of resources such as company websites, intranets, email, or existing indoctrination and recurrent training.

Accountable Executive Scalability

In smaller organizations, the accountable executive may personally participate in or directly supervise operational processes. This individual may serve in multiple positions within the company. In larger organizations, the accountable executive is responsible for ensuring that management personnel are clearly designated for ensuring the safety of operational and safety management processes.

System Description and Task Analysis Scalability

Regardless of size and complexity of organization, take the time to define your system description and task analysis is critical for success.

- **Small.** The owner/manager and/or another assigned employee(s) could perform system description and analysis. An analysis could consist of a discussion among managers such as the director of operations (DO) and/or chief pilot or other individuals designated by them.
- **Medium.** System description and analysis could be performed by a member of management or one of the designated SMS management representatives with a small workgroup of company subject matter experts (SMEs) and stakeholders.
- **Large.** System description and analysis might be performed at multiple organizational levels (corporate, division, or department) and facilitated by the Safety Department/division or its equivalent. The organization might have standing committees of SMEs and stakeholders participating at various levels.

Hazard Identification Scalability

How will your organization identify hazards, what processes will you put in place to track and document those hazards?

- **Small.** The owner/manager and/or another employee(s) could perform hazard identification as part of the system analysis.
- **Medium.** As in the system analysis, a member of management or a designated SMS management representative with a small workgroup of company SMEs and stakeholders could perform hazard identification.
- **Large.** Hazard identification might be performed at multiple organizational levels (e.g., corporate, division, or department levels) and facilitated by a safety department/division or its equivalent. The organization might have standing committees of SMEs and stakeholders participating at various levels.

Risk Analysis Scalability

How will your organization make sense of the data, how will you gather and group it into meaningful information?

- **Small.** The owner/manager and/or another employee(s) could perform risk analysis. It might be performed in conjunction (by the same individual/group) with system description and analysis, hazard identification, risk assessment, and risk control.
- **Medium.** A member of management and/or the designated management representative with a small workgroup of company SMEs and stakeholders could perform risk analysis.
- **Large.** Risk analysis might be performed at multiple organizational levels, (e.g., the corporate, division, or department levels) and facilitated by the safety department or specially trained analytical personnel shared with other departments. The organization might have standing committees of SMEs and stakeholders participating at various levels.

Risk Assessment Scalability

How will you decide what level of risk you have? Who could accept what kind of risk?

- **Small.** The owner/manager, and/or another employee(s) making the risk decisions could perform risk assessment. Risk acceptance would also probably be conducted by this individual/group. These processes could be similar to a flight risk management process, or you could use a risk matrix.

- **Medium and Large.** SRM should be coordinated across the divisional and geographic units of the company to assure integrated decision-making and communication. Decisions involving multiple systems may require joint decision making among departments or managers responsible for those systems. Many companies have standing committees made up of senior managers, who are the decision makers, supported by working groups of technical personnel. For example, the accountable executive could make company-level decisions and department managers could make the decisions for their operations. A risk matrix may be useful to determine who makes the risk

Risk Control Process Scalability

What will be your process to document the risk controls? Who will be responsible for designing and evaluating those risk controls?

- **Small.** The risk control process could be a documented activity performed by the owner/manager and/or another employee(s) designing and evaluating the risk controls. It might be performed in conjunction (by the same individual/group) with system description and analysis, hazard identification, risk analysis, and risk assessment.
- **Medium and Large.** A member of management or SMS management representatives (with a small workgroup of company SMEs and stakeholders to design the risk controls) could perform the risk control process. There would be interdepartmental coordination before the controls are implemented. After the control is approved, it is implemented and documented through the company's publication system. Implementation of risk controls may include distribution of manual revisions and training of company personnel.

Data/Information-Gathering Scalability

How much data can we collect? How much data do we want to collect? What is the process for collecting data? Once we have the data how do we track it and make sense of it to help make decisions?

- **Small.** Most of the data/information-gathering for monitoring of operational processes will likely occur as a normal business process by the management personnel who are directly involved in the day-to-day operations. For example, regularly reviewing (weekly, monthly, or quarterly) the flight dispatch logs and crewmember duty records is a form of monitoring and could be conducted during the normal course of duties.
- **Medium.** Line managers and departmental or key management personnel may observe and review day-to-day activity, noting work task inconsistencies and potential safety issues. FOQA and **Line Operations Safety Audit (LOSA)** programs may also be sources of information to monitor operations.
- **Large.** Monitoring may involve multiple levels of management, safety professionals, functional area managers (such as director of operations, director of maintenance, chief inspector, and chief pilot), trained auditors/analysts, and teams/groups of line managers. As in medium-size carriers, FOQA and LOSA programs may be employed. Operational processes may need to be coordinated across adjacent work function boundaries; effective monitoring may also need to be coordinated.

Auditing Process Scalability

What is your process to audit your organization? Who should do it?

- **Small.** The accountable executive, owner, key management person, or a trained employee as a collateral duty could carry out the auditing process periodically. Audits may also be carried out as a sub-function of normal business processes. For example, comparisons of deferred maintenance logs and repair part receipts are a form of safety auditing that is probably already accomplished routinely.
- **Medium.** In a medium-size organization, the auditing process can be accomplished by operational departmental personnel, on a periodic basis, as determined by the needs of operational decision makers.
- **Large.** In a large organization, divisional auditors typically fulfill the auditing processes on a consistent basis. Large companies may already have safety and safety/quality auditors who perform this function or, in smaller divisions, personnel from inside the divisions may perform them.

Evaluation Process Scalability

What is your process to audit your organization? Who should do it?

- **Small.** The evaluation process could be carried out periodically by the accountable executive, owner, a key management person, or designated employees as a collateral duty under the direction of the accountable executive.
- **Medium.** This process could be accomplished by the DOS or safety department on a monthly, quarterly, or other periodic basis, as determined by the information needs of the accountable executive or other senior management decision makers. Personnel resources to perform the observations and data collection for evaluations could be from a small, dedicated department or selected line personnel as a collateral duty.
- **Large.** A safety department or an **internal evaluation program (IEP)** office could accomplish evaluations on a quarterly, annual, or other periodic basis, as determined by the information needs of the Accountable Executive or other senior management decision makers. Most Part 121 companies have IEPs, and their outputs can be integrated into the SMS. Analysis of evaluations is typically performed by a safety department. The appropriate operational department would act upon the resulting data, with the safety department managing the data and assisting the responsible division(s) in resolving their issues. Most large organizations have standing management committees that consider results of evaluations and any corrective action needed.

Investigations Scalability

What is your process if there is an accident or incident? Who will be involved?

- **Small.** Investigations can be conducted by the accountable executive or assigned employees.
- **Medium.** Investigations can be conducted by a safety department, with additional assigned line personnel providing technical expertise.
- **Large.** Safety teams with specialized disciplines can conduct investigations, depending on the situation.

Employee Reporting System Scalability

How will you communicated reporting process? What will be your process to collect hazard reports?

- **Small.** An employee reporting system for a small company need not be highly sophisticated to be effective. The employees might report a hazard either orally or in a note or email to their supervisor. Collection tools could be a suggestion box, voice mail hotline, etc.
- **Medium and Large.** A medium-or large-sized company will most likely have an existing online employee reporting system or ASAPs for some employee groups. Data collection for the reporting system can take many forms, from a simple suggestion box to company websites or intranets, or a dedicated email address. Data management can be accomplished with common desktop spreadsheet or database software or specialized software.

Safety Performance Scalability

Who do you want to include as you review your safety policy and safety objectives? Who should be at regularly scheduled safety meetings?

- **Small.** The accountable executive, the department of safety (DOS), other individual managers, or other designated employees could do analysis of the data gathered as a collateral duty.
- **Medium.** Analysis of data could be done by the DOS, other individuals within a safety department, or a person(s) within each department, and then shared with other departments and management during regularly scheduled meetings.
- **Large.** Operational departments may have their own data analysis group reviewing data and analyzing the data by SMEs within the respective department, possibly supported and coordinated by a safety department.

Personnel Scalability

How many personnel should be allocated to implementing and maintaining SMS?

- **Small and Medium.** As an organization grows in size, it is normal to have additional personnel performing safety, quality, or internal evaluation functions. An SMS does not change the number and types of personnel in these situations as much as it may change the way in which these persons and organizations work and interact. For example, safety performance and assessment could be a documented activity performed by the accountable executive or a coordinated activity between the accountable executive and other operational managers, supported and coordinated by the DOS. Risk acceptance would also normally fall to managers within this group.
- **Large.** This process is best addressed at the highest level in the organization and involves the accountable executive, division vice presidents, and other defined leaders and decision makers. At each level, the company would define who is responsible for making risk acceptance decisions and what actions should be taken to either correct the problem or design new risk controls. Larger companies typically have standing management committees at the functional organization level (flight operations, technical operations/maintenance, in-flight services, dispatch/operational control) and a second body at the corporate level to assure integration, coordination, and review by the accountable executive.

Training Scalability

Organizations of all sizes may choose to either train their employees in house or to contract out the training to outside vendors. Whichever option is taken, the training must be specific to your company's SMS and operations. Who will train? How will you conduct training?

Communication Safety Considerations Scalability.

What methods are available? What methods do you want to use? Who should communicate? What is the process for communicating change?

- **Small.** Communicating safety considerations to employees will probably be simple and direct. For example, the company owner or the organization could conduct regular all-hands/employee meetings, such as "hangar talk sessions." Additionally, communication could include regular and periodic briefings to the employees, posting the status of safety issues on bulletin boards, emails to employees, and face-to-face meetings with division management teams.
- **Medium.** Communication methods may be more structured than in small organizations. Safety information may be published throughout the company by printed or electronic means. Safety meetings are likely more structured and formal. Communication and feedback may be formalized in order to provide information to individual employees as well as organization-wide information for cross-boundary issues and/or common hazards.
- **Large.** Communication is more likely to be formal and a tracking system may be used to ensure that the appropriate safety messages are delivered to the appropriate personnel. Information technology approaches, such as email broadcasts or intranet Websites, may be considered to facilitate directing the flow of safety information and recording its accomplishment for evaluation and auditing purposes.

Current Documents Scalability

Who maintains the most current documents? How do personnel know where to go? Where is the process documented?

- **Small and Medium.** The owner/manager or designee may be responsible for maintaining and distributing current versions of guidance documents. Documentation may consist of a set of typewritten documents, spreadsheets, and forms that are kept in file cabinets or binders. Managers of medium-size companies need the same type of information to make decisions, although the volume is typically larger than that of a small company and smaller than that of a large organization.
- **Large.** Documentation and recordkeeping processes for a large organization may use a WBAT, unique software applications, or development of new database tools to support risk reporting and analysis. These organizations should examine existing tools and infrastructure, as it is likely that these can be leveraged (modified) to meet SMS requirements.

Documentation of Records Scalability

Do we track meetings now? Which ones do we need to track and document? What is our records retention schedule?

- **Small.** The owner/manager or designee may be responsible for maintaining auditable records. Documentation may consist of handwritten records, spreadsheets, and completed forms that are kept in file cabinets or binders.
- **Medium.** An individual or small staff may coordinate document maintenance and retention. This staff may use a combination of paper and electronic media to administer the process. Some records may be retained by department heads in accordance with a procedure delegating this responsibility.
- **Large.** The organization may have a dedicated records staff or department whose duties include document distribution and records retention. Because of the size and complexity of the organization, the use of technology is probably more pronounced.

SUMMARY

The SMS requirements are the same regardless of the size of your organization. That said, 14 CFR Part 5 allows organizations of different sizes to meet those requirements in different ways. The SMS functions do not need to be extensive or complex to be effective. All businesses, regardless of size, may use existing systems, programs and resources to document and track safety issues to resolution.

REVIEW QUESTIONS

1. How might a Part 141 FAA pilot school be different than a large Part 121 air carrier in regard to scalability?

2. Describe the primary factors that exist between a large, medium, and small organization's SMS.

3. Compare similar organizations to your own and identify how their scale would vary.

4. What strategy will you use to get buy-in from management to provide resources for developing and maintaining of SMS?

5. What are the key ways your organization obtains data and how is it used to make fact-based decisions?

6. What about your organization makes it "complex?" How will knowing this help you establish a safety management system?

CHAPTER 6
Basic Safety Concepts

OBJECTIVES

- Understand the strengths and weaknesses of long-established approaches to safety.
- Define the key components of the Reason's Swiss cheese model.
- Explain how an organization can cause an accident.
- Describe the SHEL model.
- Summarize how the 1:600 Rule is used in your SMS program.
- Discuss how the Iceberg of Ignorance can be used create a predictive safety culture.

KEY TERMS

- 600:1 Rule
- Active Error
- Continuous Analysis and Surveillance System (CASS)
- Human Factors Analysis and Classification System (HFACS)
- Iceberg of Ignorance
- Latent Error
- Practical Drift
- Safety Thinking
- SHEL model
- SMS Safety Assurance Component
- Swiss Cheese Model
- Trans-Cockpit Authority Gradient
- Unsafe Act
- Unsafe Supervision
- Violation

THE CONCEPT OF SAFETY

Depending on the perspective, the concept of safety in aviation may have different connotations, such as:

- Zero accidents or serious incidents—a view widely held by the traveling public;
- Freedom from hazards, i.e. those factors which cause or are likely to cause harm;
- Attitudes of employees of aviation organizations towards unsafe acts and conditions;
- Error avoidance; and
- Regulatory compliance.

Whatever the connotation, they all have one underlying commonality: the possibility of absolute control. These connotations convey the idea that it would be possible—by design or intervention—to bring all variables that can precipitate bad or damaging outcomes under control in aviation operational contexts.

While the elimination of accidents and serious incidents and the achievement of absolute control are certainly desirable outcomes, they are unachievable goals in open and dynamic operational contexts. Failures and operational errors will occur in aviation, in spite of the best and most accomplished efforts to prevent them. No human activity or human-made system can be guaranteed to be absolutely free from hazards and operational errors.

Safety is therefore a concept that must encompass relatives rather than absolutes. Safety risks arising from the consequences of hazards in operational contexts must be acceptable in an inherently safe system. The key issue still resides in control, but relative control, rather than absolute control. As long as safety risks and operational errors are kept under a reasonable degree of control, a system as open and dynamic as commercial civil aviation is considered to be safe. In other words, safety risks and operational errors that are controlled to a reasonable degree are acceptable in an inherently safe system.

Safety is increasingly viewed as the outcome of the management of certain organizational processes which have the objective of keeping the safety risks of the consequences of hazards in operational contexts under organizational control. Thus, for our purposes, safety is considered to have the following meaning:

> **Safety** is the state in which the possibility of harm to persons or of property damage is reduced to, and maintained at or below, an acceptable level through a continuing process of hazard identification and safety risk management.

THE EVOLUTION OF AVIATION SAFETY

During its early years, commercial aviation was a loosely regulated activity characterized by underdeveloped technology, lack of a proper infrastructure, limited oversight, an insufficient understanding of the hazards underlying aviation operations, and production demands incommensurate with the means and resources actually available to meet such demands.

Technological improvements (due in no small measure to accident investigation), together with the eventual development of an appropriate infrastructure, led to a gradual but steady decline in the frequency of accidents, as well as an ever-increasing regulatory drive. By the 1950s, aviation was becoming one of the safest industries (in terms of accidents), as well as one of the most heavily regulated.

This new safety process focused on outcomes (accidents and/or incidents of magnitude) and relied on accident investigation to determine the cause, including the possibility, of technological failures. If technological failures were not evident, attention was turned to the possibility of rule-breaking by operational personnel.

The accident investigation would backtrack looking for a point or points in the chain of events where people directly involved in the safety breakdown did not do what they were expected to do, did something unexpected, or a combination of both. This reactive approach generated safety recommendations aimed at the specific active failure. Little emphasis was placed on the hazardous conditions that, although present, were not considered causal in the occurrence under investigation. While this perspective was quite effective in identifying *what* happened, *who* did it and *when* it happened, it was considerably less effective in disclosing why and how it happened. While at one time it was important to understand *what*, *who*, and *when*, increasingly it became necessary to understand *why* and *how* in order to fully understand safety breakdowns. In recent years, significant strides have been made in achieving this understanding. In retrospect, it is clear that aviation **safety thinking** has experienced a significant evolution over the last fifty years.

ACCIDENT CAUSATION—THE SWISS CHEESE MODEL

Sadly, the annals of aviation history are littered with accidents and tragic losses. Since the late 1950s, however, the drive to reduce the accident rate has yielded unprecedented levels of safety to a point where it is now safer to fly in a commercial airliner than to drive a car. Still, while the aviation accident rate has declined tremendously since the first flights over a century ago, the cost of aviation accidents in both lives and dollars has steadily risen. As a result, the effort to reduce the accident rate still further has taken on new meaning within both military and civilian aviation.

Why do aircraft crash? The answer is complicated. It appears that the aircrew themselves are more deadly than the aircraft they fly. In fact, it is estimated that between 70 and 80% of aviation accidents can be attributed, at least in part, to human error.

So what is it that really constitutes that 70 to 80% of human error? Some would have us believe that human error and pilot error are synonymous. Yet, simply writing off aviation accidents merely to pilot error is an overly simplistic, if not naive, approach to accident causation. After all, it is well established that accidents cannot be attributed to a single cause, or in most instances, even a single individual. In fact, even the identification of a primary cause is fraught with problems. Rather, aviation accidents are the end result of a number of causes, only the last of which are the unsafe acts of the aircrew.

The challenge for safety analysts and accident investigators alike is how best to identify and mitigate the causal sequence of events, in particular that 70 to 80% associated with human error. Additionally, human error does not only refer to errors made by the pilot. We also need to analyze human errors made by aircraft maintenance workers, flight attendants, air traffic controllers, weather forecasters, airport managers and dispatchers, among others.

A particularly appealing approach to the genesis of human error is the one proposed by James Reason in 1990, which is referred to as the **Swiss cheese model** of human error. Reason's original work involved operators of a nuclear power plant. This model has been used to analyze human errors within the aviation community. Reason describes four levels of human failure, each influencing the next (see Figure 6-1).

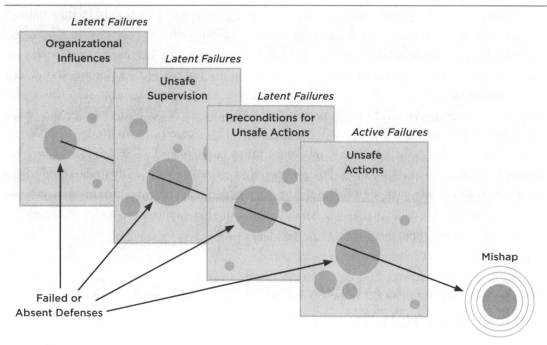

Figure 6-1   The Swiss cheese model of human error causation. *(Adapted from Reason, 1990.)*

Working backwards in time from the accident, the first level depicts those unsafe acts of operators that ultimately led to the accident. Unsafe acts reflect the active failures. More commonly referred to in aviation as aircrew/pilot error, these active failures have been the traditional focus of most accident investigations. Represented as holes in the cheese, active failures are typically the last unsafe acts committed by aircrew.

What makes the Swiss cheese model particularly useful in accident investigation is that it leads investigators to address latent failures within the causal sequence of events as well. As their name suggests, latent failures, unlike their active counterparts, may lie dormant or undetected for hours, days, weeks, or even longer, until one day they adversely affect the unsuspecting aircrew. Consequently, they are often overlooked by investigators with even the best intentions.

Within this concept of latent failures, Reason described three more levels of human failure. The first involves the condition of the aircrew as it affects performance. Referred to as preconditions for unsafe acts, this level involves conditions such as mental fatigue and poor communication and coordination practices, often referred to as crew resource management (CRM). Not surprising, if fatigued aircrew fail to communicate and coordinate their activities with others in the cockpit or individuals external to the aircraft (air traffic control or maintenance, for instance), poor decisions are made and errors often result.

But exactly why did communication and coordination break down in the first place? This is perhaps where Reason's work departed from more traditional approaches to human error. In many instances, the breakdown in good CRM practices can be traced back to instances of unsafe supervision, the third level of human failure. If, for example, two inexperienced (and perhaps even below average pilots) are paired with each other and sent on a flight into known adverse weather at night, is anyone really surprised by a tragic outcome? To make matters worse, if this questionable manning practice is coupled with the lack of quality CRM training, the potential for miscommunication and ultimately, aircrew errors, is magnified. In a sense then, the crew was set up for failure as crew coordination and ultimately performance would be compromised. This is not to lessen the role played by the aircrew, only that intervention and mitigation strategies might lie higher within the system.

Reason's model didn't stop at the supervisory level either; the organization itself can impact performance at all levels. For instance, in times of fiscal austerity, funding is often cut, and as a result, training and flight time are curtailed. Consequently, supervisors are often left with no alternative but to task non-proficient aviators with complex tasks. Not surprisingly then, in the absence of good CRM training, communication and coordination failures will begin to appear as will a myriad of other preconditions, all of which will affect performance and elicit aircrew errors. Therefore, it makes sense that, if the accident rate is going to be reduced beyond current levels, investigators and analysts alike must examine the accident sequence in its entirety and expand it beyond the cockpit. Ultimately, causal factors at all levels within the organization must be addressed if any accident investigation and prevention system is going to succeed. In many ways, Reason's Swiss cheese model has revolutionized common views of accident causation.

Unsafe Acts

The unsafe acts of aircrew can be loosely classified into two categories: errors and violations. In general, errors represent the mental or physical activities of individuals that fail to achieve their intended outcome. Not surprising, given the fact that human beings by their very nature make errors, these unsafe acts dominate most accident databases. Violations, on the other hand, refer to the willful disregard for the rules and regulations that govern the safety of flight. Still, distinguishing between errors and violations does not provide the level of granularity required of most accident investigations. Therefore, the categories of errors and violations were expanded here (see Figure 6-2), as elsewhere, to include three basic error types (skill-based, decision, and perceptual) and two forms of violations (routine and exceptional).

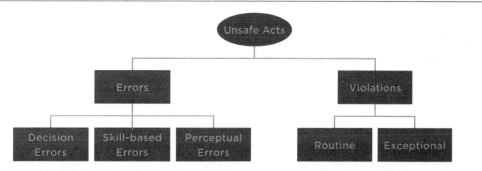

Figure 6-2 Categories of unsafe acts committed by aircrews.

Violations

By definition, errors occur within the rules and regulations espoused by an organization; typically dominating most accident databases. In contrast, violations represent a willful disregard for the rules and regulations that govern safe flight. Fortunately, these occur much less frequently, however, they often involve fatalities.

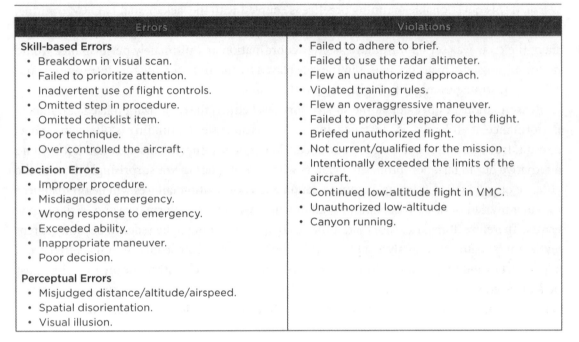

Errors	Violations
Skill-based Errors	• Failed to adhere to brief.
• Breakdown in visual scan.	• Failed to use the radar altimeter.
• Failed to prioritize attention.	• Flew an unauthorized approach.
• Inadvertent use of flight controls.	• Violated training rules.
• Omitted step in procedure.	• Flew an overaggressive maneuver.
• Omitted checklist item.	• Failed to properly prepare for the flight.
• Poor technique.	• Briefed unauthorized flight.
• Over controlled the aircraft.	• Not current/qualified for the mission.
	• Intentionally exceeded the limits of the aircraft.
Decision Errors	• Continued low-altitude flight in VMC.
• Improper procedure.	• Unauthorized low-altitude
• Misdiagnosed emergency.	• Canyon running.
• Wrong response to emergency.	
• Exceeded ability.	
• Inappropriate maneuver.	
• Poor decision.	
Perceptual Errors	
• Misjudged distance/altitude/airspeed.	
• Spatial disorientation.	
• Visual illusion.	

Table 6-1 Selected examples of unsafe acts of pilot operators. *(Note: not a complete list.)*

While there are many ways to distinguish between types of violations, two distinct forms have been identified, based on their etiology, that will help the safety professional when identifying accident causal factors. The first, routine violations, tend to be habitual by nature and often tolerated by governing authority. Consider, for example, the individual who drives consistently 5-10 mph faster than allowed by law or someone who routinely flies in marginal weather when authorized for visual meteorological conditions only. While both are certainly against the governing regulations, many others do the same thing. Furthermore, individuals who drive 64 mph in a 55-mph zone, almost always drive 64 in a 55-mph zone. That is, they routinely violate the speed limit. The same can typically be said of the pilot who routinely flies into marginal weather.

What makes matters worse, these violations (commonly referred to as "bending the rules") are often known and tolerated and, in effect, sanctioned by supervisory authority (i.e., you're not likely to get a traffic citation until you exceed the posted speed limit by more than 10 mph). If, however, the local authorities started handing out traffic citations for exceeding the speed limit on the highway by nine mph or less then it is less likely that individuals would violate the rules. Therefore, by definition, if a routine violation is identified, one must look further up the supervisory chain to identify those individuals in authority who are not enforcing the rules.

On the other hand, unlike routine violations, exceptional violations appear as isolated departures from authority, neither necessarily indicative of individual's typical behavior pattern nor condoned by management. For example, an isolated instance of driving 105 mph in a 55 mph zone is considered an exceptional violation. Likewise, flying under a bridge or engaging in other prohibited maneuvers, like low-level canyon running, would constitute an exceptional violation. It is important to note that, while most exceptional violations are appalling, they are not considered exceptional because of their extreme nature. Rather they are considered exceptional because they are neither typical of the individual nor condoned by authority. Still, what makes exceptional violations particularly difficult for any organization to deal with is that they are not indicative of an individual's behavioral repertoire and, as such, are particularly difficult to predict. In fact, when individuals are confronted with evidence of their dreadful behavior and asked to explain it, they are often left with little explanation. Those individuals who survived such excursions from the norm clearly knew that, if caught, dire consequences would follow. Defying all logic, many otherwise model citizens have been down this potentially tragic road.

Preconditions for Unsafe Acts

Arguably, the unsafe acts of pilots can be directly linked to nearly 70 to 80% of all aviation accidents. Yet simply focusing on unsafe acts is like focusing on a fever without understanding the underlying disease causing it. Investigators must dig deeper into why the unsafe acts took place. As a first step, two major subdivisions of unsafe aircrew conditions were developed: substandard conditions of operators and the substandard practices they commit (see Figure 6-3 below and Table 6-2 on the next page).

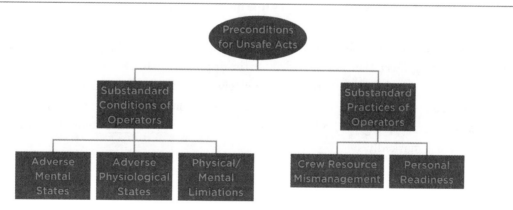

Figure 6-3   Categories of preconditions of unsafe acts.

Substandard Conditions of Operators	Substandard Practice of Operators
Adverse Mental States • Channelized attention. • Complacency. • Distraction. • Mental Fatigue. • Get-home-itis. • Haste. • Loss of situational awareness. • Misplaced motivation. • Task saturation. **Adverse Physiological States** • Impaired physiological state. • Medical illness. • Physiological incapacitation. • Physical fatigue. **Physical/Mental Limitation** • Insufficient reaction time. • Visual limitation. • Incompatible intelligence/aptitude. • Incompatible physical capability.	**Crew Resource Management** • Failed to back-up. • Failed to communicate/coordinate. • Failed to conduct adequate brief. • Failed to use all available resources. • Failure of leadership. • Misinterpretation of traffic calls. **Personal Readiness** • Excessive physical training. • Self-medicating. • Violation of crew rest requirement. • Violation of bottle-to-throttle requirement.

Table 6-2   Selected examples of unsafe aircrew conditions. *(Note: not a complete list.)*

Inadequate Supervision

The role of any supervisor is to provide the opportunity to succeed. To do this, the supervisor, no matter at what level of operation, must provide guidance, training opportunities, leadership, and motivation, as well as the proper role model to be emulated. Unfortunately, this is not always the case.

For example, it is not difficult to conceive of a situation where adequate crew resource management training was either not provided, or the opportunity to attend such training was not afforded to a particular aircrew member. Conceivably, aircrew coordination skills would be compromised and if the aircraft were put into an adverse situation (an emergency for instance), the risk of an error being committed would be exacerbated and the potential for an accident would increase markedly.

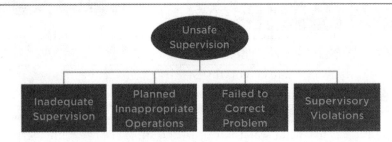

Figure 6-4   Categories of unsafe supervision.

In a similar vein, sound professional guidance and oversight is an essential ingredient of any successful organization. While empowering individuals to make decisions and function independently is certainly essential, this does not divorce the supervisor from accountability. The lack of guidance and oversight has proven to be the breeding ground for many of the violations that have crept into the cockpit. As such, any thorough investigation of accident causal factors must consider the role supervision plays (such as whether the supervision was inappropriate or did not occur at all) in the genesis of human error (Table 6-3).

Inadequate Supervision	Failure to Correct a Known Problem
• Failed to provide guidance. • Failed to provide operational doctrine. • Failed to provide oversight. • Failed to provide training. • Failed to track qualifications. • Failed to track performance. **Planned Inappropriate Operations** • Failed to provide correct data. • Failed to provide adequate brief time. • Improper manning. • Mission not in accordance with rules/ regulations. • Provided inadequate opportunity for crew rest.	• Failed to correct document in error. • Failed to identify an at-risk aviator. • Failed to initiate corrective action. • Failed to report unsafe tendencies. **Supervisory Violations** • Authorized unnecessary hazard. • Failed to enforce rules and regulations. • Authorized unqualified crew for flight.

Table 6-3   Selected examples of unsafe supervision. *(Note: not a complete list.)*

Planned Inappropriate Operations

Occasionally, the operational tempo and the scheduling of aircrew are such that individuals are put at unacceptable risk, crew rest is jeopardized, and ultimately performance is adversely affected. Such operations, though arguably unavoidable during emergencies, are unacceptable during normal operations. Therefore, the second category of unsafe supervision, planned inappropriate operations, was created to account for these failures (Table 6-3).

Take, for example, the issue of improper crew pairing. It is well known that when very senior, dictatorial captains are paired with very junior, weak co-pilots, communication and coordination problems are likely to occur. Commonly referred to as the **trans-cockpit authority gradient**, such conditions likely contributed to the tragic crash of a commercial airliner into the Potomac River outside of Washington, DC, in January of 1982. In that accident, the captain of the aircraft repeatedly rebuffed the first officer when the latter indicated that the engine instruments did not appear normal. Undaunted, the captain continued a fatal takeoff in icing conditions with less than adequate takeoff thrust. The aircraft stalled and plummeted into the icy river, killing the crew and many of the passengers.

Clearly, the captain and crew were held accountable. They died in the accident and cannot shed light on causation; but what was the role of the supervisory chain? Perhaps crew pairing was equally responsible. Although not specifically addressed in the report, such issues are clearly worth exploring in many accidents. In fact, in that particular accident, several other training and manning issues were identified.

Failure to Correct a Known Problem

The third category of known unsafe supervision, failed to correct a known problem, refers to those instances when deficiencies among individuals, equipment, training or other related safety areas are known to the supervisor, yet are allowed to continue unabated (Table 6-3). For example, it is not uncommon for accident investigators to interview the pilot's friends, colleagues, and supervisors after a fatal crash only to find out that they "knew it would happen to him some day." If the supervisor knew that a pilot was incapable of flying safely, and allowed the flight anyway, he clearly did the pilot no favors. The failure to correct the behavior, either through remedial training or, if necessary, removal from flight status, essentially signed the pilot's death warrant—not to mention that of others who may have been on board.

Likewise, the failure to consistently correct or discipline inappropriate behavior certainly fosters an unsafe atmosphere and promotes the violation of rules. Aviation history is rich with by reports of aviators who tell hair-raising stories of their exploits and barnstorming low-level flights (the infamous "been there, done that"). While entertaining to some, they often serve to promulgate a perception of tolerance and "one-up-manship" until one day someone ties the low altitude flight record of ground-level! Indeed, the failure to report these unsafe tendencies and initiate corrective actions is yet another example of the failure to correct known problems.

Supervisory Violations

Supervisory violations, on the other hand, are reserved for those instances when existing rules and regulations are willfully disregarded by supervisors (Table 6-3). Although arguably rare, supervisors have been known occasionally to violate the rules and doctrine when managing their assets. For instance, there have been occasions when individuals were permitted to operate an aircraft without current qualifications or license. Likewise, it can be argued that failing to enforce existing rules and regulations or flaunting authority are also violations at the supervisory level. While rare and possibly difficult to cull out, such practices are a flagrant violation of the rules and invariably set the stage for the tragic sequence of events that predictably follow.

Organizational Influences

As noted previously, fallible decisions of upper-level management directly affect supervisory practices, as well as the conditions and actions of operators. Unfortunately, these organizational errors often go unnoticed by safety professionals, due in large part to the lack of a clear framework from which to investigate them. Generally speaking, the most elusive of latent failures revolve around issues related to resource management, organizational climate, and operational processes, as detailed below in Figure 6-5.

Figure 6-5   Organizational factors influencing accidents.

Resource Management

This category encompasses the realm of corporate-level decision making regarding the allocation and maintenance of organizational assets such as human resources (personnel), monetary assets, and equipment/facilities (Table 6-4). Generally, corporate decisions about how such resources should be managed center around two distinct objectives—the goal of safety and the goal of on-time, cost-effective operations. In times of prosperity, both objectives can be easily balanced and satisfied in full. However, as we mentioned earlier, there may also be times of fiscal austerity that demand some give and take between the two. Unfortunately, history tells us that safety is often the loser in such battles and, as some can attest to, safety and training are often the first to be cut in organizations having financial difficulties. If cutbacks in such areas are too severe, flight proficiency may suffer, and the best pilots may leave the organization for greener pastures.

Resource/Acquisition Management	Organizational Process

• Human resources
  ▷ Selection
  ▷ Staffing/manning
  ▷ Training
• Monetary/budget resources
  ▷ Excessive cost cutting
  ▷ Lack of funding
• Equipment/facility resources
  ▷ Poor design
  ▷ Purchasing of unsuitable equipment
 	
Organizational Climate
• Structure
  ▷ Chain-of-command
  ▷ Delegation of authority
  ▷ Communication
  ▷ Formal accountability for actions
• Policies
  ▷ Hiring and firing
  ▷ Promotion
  ▷ Drugs and alcohol
• Culture
  ▷ Norms and rules
  ▷ Values and beliefs
  ▷ Organizational justice
	• Operations
	  ▷ Operational tempo
	  ▷ Time pressure
	  ▷ Production quotas
	  ▷ Incentives
	  ▷ Measurement/appraisal
	  ▷ Schedules
	  ▷ Deficient planning
	• Procedures
	  ▷ Standards
	  ▷ Clearly defined objectives
	  ▷ Documentation
	  ▷ Instructions
	• Oversight
	  ▷ Risk management
	  ▷ Safety programs

Table 6-4 Selected examples of organizational influences. *(Note: not a complete list.)*

Excessive cost cutting could also result in reduced funding for new equipment or may lead to the purchase of equipment that is sub optimal and inadequately designed for the type of operations flown by the company. Other trickle-down effects include poorly maintained equipment and workspaces, and the failure to correct known design flaws in existing equipment. The result is a scenario involving unseasoned, less-skilled pilots flying old and poorly maintained aircraft under the least desirable conditions and schedules. The ramifications for aviation safety are not hard to imagine.

Culture and Climate

The organizational culture or climate refers to a broad class of organizational variables that influence worker performance. Formally, the terms culture and climate refer to the same thing. An organizational culture can be viewed as the working atmosphere within the organization. One telltale sign of an organization's culture is its structure, as reflected in the chain of command, delegation of authority and responsibility, communication channels, and formal accountability for actions (Table 6-4). Just like in the cockpit, communication and coordination are vital within an organization. If management and staff within an organization are not communicating, or if no one knows who is in charge, organizational safety clearly suffers and accidents do happen.

An organization's policies are also good indicators of its culture. Policies are official guidelines that direct management's decisions about such things as hiring and firing, promotion, retention, raises, sick leave, drugs and alcohol, overtime, accident investigations, and the use of safety equipment. Culture, on the other hand, refers to the unofficial or unspoken rules, values, attitudes, beliefs, and customs of an organization. Culture is "the way things really get done around here."

When policies are ill defined, adversarial, or conflicting, or when they are supplanted by unofficial rules and values, confusion abounds within the organization. Indeed, there are some corporate managers who are quick to give "lip service" to official safety policies while in a public forum, but then overlook such policies when operating behind the scenes. Safety is bound to suffer under such conditions.

Operational Process

This category refers to corporate decisions and rules that govern the everyday activities within an organization, including the establishment and use of standardized operating procedures and formal methods for maintaining checks and balances (oversight) between the workforce and management. For example, such factors as operational tempo, time pressures, incentive systems, and work schedules are all factors that can adversely affect safety (Table 6-4). As stated earlier, there may be instances when those within the upper echelon of an organization determine that it is necessary to increase the operational tempo to a point that overextends a supervisor's staffing capabilities. Therefore, a supervisor may resort to the use of inadequate scheduling procedures that jeopardize crew rest and produce sub optimal crew pairings, putting aircrew at an increased risk of a mishap. Organizations should have official procedures in place to address such contingencies, as well as oversight programs to monitor such risks.

The **Human Factors Analysis and Classification System (HFACS)** framework provides a comprehensive, user-friendly tool for identifying and classifying the human causes of aviation accidents. The system, which is based upon Reason's Swiss cheese model of latent and active failures, encompasses all aspects of human error, including the conditions of operators and organizational failure. Still, HFACS and any other framework only contributes to an already burgeoning list of human error taxonomies if it does not prove useful in the operational setting.

THE 1:600 RULE

Research into industrial safety accidents in 1969 indicated that for every 600 reported occurrences with no injury or damage, there were:

- 30 incidents involving property damage;
- 10 accidents involving serious injuries; and
- One major fatal injury.

The 1-10-30-600 ratio shown in Figure 6-6 is indicative of a wasted opportunity if investigative efforts are focused only on those rare occurrences where there is serious injury or significant damage. The factors contributing to such accidents may be present in hundreds of incidents and could be identified. The root cause could be identified before serious injury or damage ensures. Effective safety management requires that staff and management identify and analyze hazards before they result in accidents.

Figure 6-6   The 1:600 rule.

THE ICEBERG OF IGNORANCE

In his acclaimed study The Iceberg of Ignorance, consultant Sidney Yoshida concluded: "Only 4% of an organization's front line problems are known by top management, 9% are known by middle management, 74% by supervisors and 100% by employees." (See Figure 6-7.)

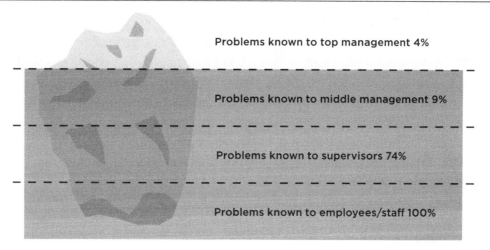

Figure 6-7   The iceberg of ignorance.

Melting the Ignorance in Your Organization

Top-level management that make strategic plans, decide on budgetary allocations, review benefit plans and such need all the input they can get to make the best, informed decisions. This business-like approach can also be used to instill a predictive type of safety management. Top-level management can learn a lot from asking their subordinates "Where will we have our next accident?"

THE SHELL MODEL

One needs to understand how the various components and features of the operational context and the interrelationships between components, features and people may affect human operational performance. A simple, yet visually powerful, conceptual tool for the analysis of the components and features of operational contexts and their possible interactions with people is the **SHEL model**. The SHEL model (sometimes referred to as

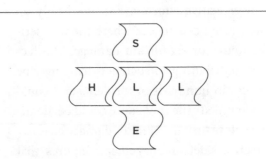

Figure 6-8 The SHEL model.

the SHEL(L) model) can be used to help visualize the interrelationships among the various components and features of the aviation system. This model places emphasis on the individual and the human's interfaces with the other components and features of the aviation system. The SHEL model's name is derived from the initial letters of its four components:

- Software (S) (procedures, training, support);
- Hardware (H) (machines and equipment);
- Environment (E) (the operating circumstances in which the rest of the L-H-S system must function); and
- Liveware (L) (humans in the workplace).

Figure 6-8 depicts the SHEL model. This building-block diagram is intended to provide a basic understanding of the relationship of individuals to components and features in the workplace.

Liveware

In the center of the SHEL model are the humans at the front line of operations. Although humans are remarkably adaptable, they are subject to considerable variations in performance. Humans are not standardized to the same degree as hardware, so the edges of this block are not simple and straight. Humans do not interface perfectly with the various components of the world in which they work. To avoid tensions that may compromise human performance, the effects of irregularities at the interfaces between the various SHEL blocks and the central Liveware block must be understood. The other components of the system must be carefully matched to humans if stresses in the system are to be avoided.

Several different factors put the rough edges on the Liveware block. Some of the more important factors affecting individual performance are:

- **Physical factors.** These include the human's physical capabilities to perform the required tasks, such as strength, height, reach, vision and hearing.
- **Physiological factors.** These include those factors which affect the human's internal physical processes, which can compromise physical and cognitive performance, including oxygen availability, general health and fitness, disease or illness, tobacco, drug or alcohol use, personal stress, fatigue, and pregnancy.
- **Psychological factors.** These include those factors affecting the psychological preparedness of the human to meet all the circumstances that might occur, such as adequacy of training, knowledge and experience, and workload.
- **Psycho-social factors.** These include all those external factors in the social system of humans that bring pressure to bear on them in their work and non-work environments; for example: an argument with a supervisor, labor-management disputes, a death in the family, personal financial problems or other domestic tension.

The SHEL model is particularly useful in visualizing the interfaces between the various components of the aviation system. These include:

- **Liveware-Hardware (L-H).** The interface between the human and technology is the one most commonly considered when speaking of human performance. It determines how the human interfaces with the physical work environment, such as the design of seats to fit the sitting characteristics of the human body, displays to match the sensory and information processing characteristics of the user, and proper movement, coding, and location of controls for the user. There is a natural human tendency to adapt to L-H mismatches. This tendency may mask serious deficiencies, which may only become evident after an occurrence.
- **Liveware-Software (L-S).** The L-S interface is the relationship between the human and the supporting systems found in the workplace, encompassing regulations, manuals, checklists, publications, standard operating procedures (SOPs) and computer software. It includes such "user-friendliness" issues as currency, accuracy, format and presentation, vocabulary, clarity, and symbology.
- **Liveware-Liveware (L-L).** The L-L interface is the relationship between the human and other persons in the workplace. Flight crews, air traffic controllers, aircraft maintenance engineers and other operational personnel function as groups, and group influences play a role in determining human performance. The advent of crew resource management (CRM) has resulted in considerable focus on this interface. CRM training and its extension to air traffic services (ATS) (team resource management (TRM)) and maintenance (maintenance resource management (MRM)) focus on the management of operational errors. Staff/management relationships are also within the scope of this interface, as are corporate culture, corporate climate and company operating pressures, which can all significantly affect human performance.
- **Liveware-Environment (L-E).** This interface involves the relationship between the human and both the internal and external environments. The internal workplace environment includes such physical considerations as temperature, ambient light, noise, vibration and air quality. The external environment includes such things as visibility, turbulence, and terrain. The twenty-four hour a day, seven days a week, aviation work

environment includes disturbances to normal biological rhythms that include sleep patterns. In addition, the aviation system operates within a context of broad political and economic constraints, which in turn affect the overall corporate environment. Included here are such factors as the adequacy of physical facilities and supporting infrastructure, the local financial situation, and regulatory effectiveness. Just as the immediate work environment may create pressures to take short cuts, inadequate infrastructure support may also compromise the quality of decision-making.

Care needs to be taken in order that operational errors do not filter through the cracks at the interfaces. For the most part, the rough edges of these interfaces can be managed, for example:

- The designer can ensure the performance reliability of the equipment under specified operating conditions.
- During the certification process, the regulatory authority can define realistic conditions under which the equipment may be used.
- The organization's management can develop standard operations procedures (SOPs) and provide initial and recurrent training for the safe use of the equipment.
- Individual equipment operators can ensure their familiarity and confidence in using the equipment safely under all required operating conditions.

PRACTICAL DRIFT

The ICAO Safety Management System Manual discusses the theory of practical drift and how it relates to safety. Practical drift explains how the baseline performance of any system drifts away from its original design when the organization's processes and procedures cannot anticipate all situations that may arise in daily operations.

During the early stages of system design operational interactions between people and technology, as well as the operational context, are taken into consideration to identify the expected performance limitations as well as potential hazards. The initial system design is based on three fundamental assumptions:

- The technology needed to achieve the system production goals is available,
- The people are trained to properly operate the technology, and
- The regulations and procedures will dictate system and human behavior.

These assumptions underlie the baseline (or ideal) system performance, which can be graphically presented as a straight line from the date of operational deployment until the system is decommissioned (see Figure 6-9). Once operationally deployed, the system performs as designed, following baseline performance most of the time. In reality, however, operational performance is different from baseline performance as a consequence of real-life operations and changes in the operational and regulatory environment. Since the drift is a consequence of daily practice, it is referred to as a practical drift. The term practical drift is used in this context as the gradual departure from an intended course because of external influences.

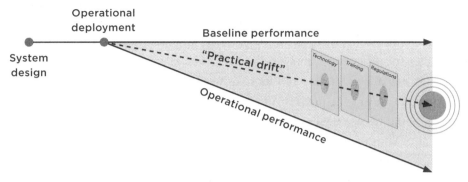

Figure 6-9   Practical drift. *(Scott A. Snook)*

A practical drift from baseline performance to operational performance is foreseeable in any system, no matter how careful and well thought out its design planning may have been. Capturing and analyzing the information on what takes place within the practical drift holds considerable learning potential about successful safety adaptations and, therefore, for the control and mitigation of safety risks. The closer to the beginning of the practical drift that the information can be systematically captured, the greater the number of hazards and safety risks that can be predicted and addressed, leading to formal interventions for re-design of or improvements to the system. However, the unchecked proliferation of local adaptations and personal strategies may lead the practical drift to depart too far from the expected baseline performance, to the extent that an incident or an accident becomes a greater possibility.

EFFECTIVE SAFETY PROGRAM MANAGEMENT

The commercial air carrier accident rate in the United States has decreased substantially over the past 10 years. This has been accomplished through a growing body of regulations, FAA oversight activities, and voluntary industry safety initiatives.

During the past 10 years, however, the FAA has identified a more recent trend involving hazards that were revealed during incident and accident investigations. Many of these hazards could have been mitigated or eliminated earlier had a structured, organization-wide approach to managing air carrier's operations been in place. For example, FAA's Office of Accident Investigation and Prevention identified 172 accidents involving 14 CFR Part 121 operators from fiscal year (FY) 2001 through FY 2010 that could have been mitigated if air carriers had implemented a safety management system to identify hazards in their daily operations and developed methods to control the risk. The following two accidents are representative of the 172 accidents reviewed by the FAA and discussed in the Initial Regulatory Evaluation. Summaries of these two accidents are included to illustrate the potential mitigations that could have resulted with SMS.

On January 8, 2003, Air Midwest flight 5481 crashed immediately after lift-off in Charlotte, North Carolina. The aircraft was destroyed by impact and post impact fire, resulting in 21 fatalities and one injury to a person on the ground. This accident occurred shortly after out-sourced maintenance was completed on the airplane's elevator control system. The accident investigation revealed that the elevator controls were improperly rigged during maintenance. The crew was not aware of this unsafe condition. The following is an example of how mainte-nance hazards could have been identified and their associated risks mitigated if the carrier had implemented an SMS.

A safety risk management analysis would have been triggered by the air carrier's plan to have aircraft maintenance performed at uncertificated repair facility using maintenance tech-nicians provided by a third party sub-contractor. First, the air carrier's maintenance manage-ment would have conducted a thorough system analysis, reviewing its current maintenance program, including all relevant policies, processes, and procedures. It would have identified the personnel, procedures, equipment, and facilities necessary to perform the work and assessed whether the maintenance facility, its management, and the third party mechanics met those requirements. It also would have identified the personnel necessary to conduct oversight for the air carrier at the maintenance facility. Following the system analysis, the air carrier's mainte-nance management would have identified the following system hazards:

- The maintenance facility was not a certificated repair station and therefore lacked the controls associated with regulatory certification.
- The facility, its management and the actual workforce were provided by separate con-tractors.
- The inadequate number of experienced air carrier maintenance representatives and their lack of authority under the contract to oversee the performance of the maintenance. The maintenance management team would have reported these issues to the management representative and the accountable executive.

The air carrier's maintenance management, in assessing the risk of these and other hazards, would have considered the worst credible outcome of the performance of the maintenance at that facility under those conditions. Those risks may have been determined to be unacceptable and appropriate risk controls would have been implemented. Such risk control options may have included contracting with a certificated Part 145 repair station, revising the maintenance procedures and associated job aids for its maintenance and inspection programs, having ad-ditional experienced maintenance representatives of the air carrier, with appropriate contract authorities stationed at the repair facility to monitor the performance of maintenance tasks and inspections.

Also, through the SMS safety assurance processes, the air carrier would have evaluated the safety performance of its risk controls through its **continuous analysis and surveillance system (CASS)** to verify that the controls were effective. Errors in specific maintenance tasks or inspections may have been spotted by the onsite air carrier maintenance representatives or through a confidential employee reporting system if any of these concerns were raised with regard to the maintenance activities. These reports would have been utilized to steer changes in existing policies or in more effective contracting and execution of maintenance.

Using the SMS safety promotion component, the air carrier could have made these critical maintenance issues known to its entire maintenance workforce, including air carrier management. This would have increased awareness of hazards and enhanced the safety of the overall maintenance program for the air carrier.

A second example is Comair flight 5191. On August 27, 2006, at approximately 6 a.m., Comair flight 5191 crashed during takeoff from Blue Grass Airport, Lexington, Kentucky, enroute to Atlanta, Georgia. The flightcrew received and acknowledged a clearance from the tower to take off from runway 22, but instead they positioned the airplane on runway 26 and commenced the takeoff. The airplane ran off the end of the runway and impacted the airport perimeter fence, trees, and terrain. The pilot-in-command (PIC), flight attendant, and 47 passengers were killed. The second-in-command pilot sustained serious injuries. The airplane was destroyed by impact forces and a post-crash fire. The flightcrew believed that they had taxied the airplane to runway 22 when they had actually taxied onto Runway 26 and initiated the takeoff roll. The flightcrew's noncompliance with standard operating procedures, including the PIC's abbreviated taxi briefing, combined with both pilots' non-pertinent conversation most likely created an atmosphere in the cockpit that enabled the crew's errors. This is an example of how hazards relating to the flight operations of this accident could have been identified and the associated risks mitigated if the carrier had implemented an SMS.

In this instance, the **SMS safety assurance component** would have triggered a formal safety risk management analysis. Under the SMS safety assurance process, periodic audits of flight crew performance, such as LOSA, may have revealed systemic failures of crew coordination concepts and failures to follow standard procedures. Additionally, reports from a confidential employee reporting system like ASAP would have indicated that deficiencies in flightcrew performance. LOSA or other structured operational checking procedures, combined with reports from a confidential employee reporting system regarding flight crew performance, would have indicated that the existing controls, such as operational procedures and preflight checklists were not effective, or flightcrew training and evaluation programs were ineffective.

Under a formal SMS safety risk management process, the management representative would have ensured that the flight operations management team conducted a system analysis, reviewing its operational control and flight operations procedures, the operating environment (runway conditions, airport configuration), as well as the personnel and equipment required for the safe operation of the airplane. The system analysis would have led to a discovery of hazards and possible errors that could be made at runway intersections, like the incorrect selection of the appropriate departure runway. The flight operations management team would have reported these issues to the management representative and the accountable executive.

Upon completion of the risk assessment, the flight operations management team could have developed risk controls, such as revising the checklists to require the positive verification of the airplane alignment on the correct runway and additional crew resource management training to enhance the crewmembers' situational awareness. These procedures could be incorporated into the company's flight manuals, checklists, and training curriculum. Once in place, the effectiveness of the risk controls would have been continuously monitored under the safety assurance processes.

From the SMS safety promotion component, the information gained through the safety risk management and safety assurance processes such as the employee reporting system, could be provided back to crews in the form of awareness tools such as company newsletters, bulletins to pilots, and other communications media.

While the elimination of aircraft accidents and/or serious incidents remains the ultimate goal, it is recognized that the aviation system cannot be completely free of hazards and associated risks. Human activities or human-built systems cannot be guaranteed to be absolutely free from operational errors and their consequences. Therefore, safety is a dynamic characteristic of the aviation system, whereby safety risks must be continuously mitigated. It is important to note that the acceptability of safety performance is often influenced by domestic and international norms and culture. As long as safety risks are kept under an appropriate level of control, a system as open and dynamic as aviation can still be managed to maintain the appropriate balance between production and protection.

Safety Program and SMS are Complimentary Functions

A safety program and an SMS are complimentary functions within the management system. Both provide top management feedback regarding the health and effectiveness of the organization as a whole.

- Safety programs include the capture and analysis of employee safety/hazard reports, the investigation of operational incidents and accidents, the oversight of risk assessment activities, and flight operations data assessment programs. As commonly practiced, safety programs are often reactive in nature in that they involve analysis of events that have already occurred (investigative efforts are oriented to identify root cause and establish corrective actions to prevent reoccurrence or limit frequency to acceptable risk levels).

- In contrast, an SMS is designed to systematically and proactively search for weaknesses in the management system, operational processes/procedures, or documentation. An effective IEP seeks to assure that key processes and controls are in place across the full spectrum of operational safety. The IEP is focused on verifying organizational compliance with all external regulatory requirements and internal organizational policies and procedures. These comprehensive system audits identify opportunities for improvement, which ultimately enhance safety through establishment of predictable and reliable business processes.

Managing Safety Within Your SMS Program

As aviation operational procedures and equipment (hardware and software) evolve, their interaction and interdependency must be addressed. Within the system of your organization, the discovery and mitigation of a safety hazard often falls within the purview of multiple specialties within your organization.

The effects of safety hazards and associated mitigations across multiple organizations, domains, and implementation timelines must be properly understood to achieve the highest practical level of safety. Safety risk deemed acceptable for an individual element of the NAS may lead to unintentional safety risk in another if a safety assessment is not conducted with a "system of systems" philosophy. As emerging NAS equipment, operations, and procedures are tested and implemented, safety risk assessments must account for their potential safety impact on existing/legacy tools and procedures, and vice versa. Sharing safety data and conducting co-operative analyses using an integrated safety management approach helps identify and resolve issues requiring the consideration of multiple disciplines.

The goal of an integrated approach to safety management is to eliminate gaps in safety analyses by assessing NAS equipment, operations, and procedures across three planes: vertical, horizontal, and temporal. The vertical plane is hierarchical, providing assessments from a specific project up to the NAS-level system of systems of which the project is a part. The horizontal plane spans organizations, programs, and systems. Finally, the temporal plane attempts to eliminate safety gaps across program and system implementation timelines.

REVIEW QUESTIONS

1. Using Reason's Swiss cheese model, identify the correct root cause of the following accidents.
   a. The noise and vibration in the cockpit made it difficult to hear and understand speech. Physical environment.
   b. Fitness for duty.
   c. Technological environment.

Using Reason's Swiss cheese model (page 71), identify the correct root cause of the following accidents in questions 2 through 4.

2. Against established policy and management direction, the pilot taxied up to the gate without any ramp personnel to guide him.
   a. Decision error.
   b. Routine violation.
   c. Exceptional violation.

3. Though the boss knows that a particular pilot is a risk-taker and sometimes pushed the envelope when flying, he did not take action (e.g. by confronting the pilot about his way of flying).
   a. Failure to correct a known problem.
   b. Routine error.
   c. Unsafe act.

4. The pilot misunderstood the instructions from air traffic control and did not seek clarification.
   a. Communication and coordination.
   b. Physical limitation.
   c. All the above.

5. What percentage of aviation accidents can be attributed, in part, to human error?
   a. Approximately 25-50%.
   b. Approximately 70-80%.
   c. Approximately 50-75%.

6. Explain how an organization can cause an accident.

7. Describe the key components of the SHEL Model and how they relate to your SMS.

8. Summarize how the 1:600 Rule is used in your SMS Program.

9. Discuss how the Iceberg of Ignorance can be used create a predictive safety culture.

CHAPTER 7
SMS Planning and Process

OBJECTIVES

- To identify and describe the steps necessary to be recognized by the FAA and obtain SMS active conformance recognition through the SMS Voluntary Program.
- To identify the steps necessary to be recognized by the state and obtain SMS compliance through and 14 CFR Part 5 and AC 120-92 for air carriers operating under Part 121 rules.
- To develop an appreciation for the complexity of an SMS, including the time and financial commitment necessary to create and implement one.
- Explore the various resources available to develop greater understanding of the process to reach active applicant, active participant, and active conformance under the SMS Voluntary Program.

KEY TERMS

- Active Applicant
- Active Conformance
- Active Participant
- Certificate Holder (Service Provider/Organization)
- Continued Operational Safety (COS)
- Implementation Support Team
- Schedule of Events
- SMS Implementation Plan
- SMS Pilot Program
- SMS Program Office (SMSPO)
- SMSVP Guide
- SMSVP Non Active Participant
- SMSVP Standard
- Validation Project Plan (VPP)

INTRODUCTION

By March 9, 2018, 14 CFR Part 5 requires certificate holders authorized to conduct operations under Part 121 to develop, implement, and be in SMS compliance. For SMS to be considered implemented and complied with, a certificate holder must have SMS documentation and resources (hardware, software, personnel) in place, often referred to as "design," and be able to demonstrate the performance of their intended functions. Adhering to Part 5 U.S. SMS requirements for air carriers operating under Part 121 brings the U.S. in harmony with ICAO SMS standards.

Simultaneously, aviation service providers who do not operate under Part 121 may desire to meet FAA and ICAO SMS requirements, but are not required to do so by FAA regulations. These Aviation Service Providers or Non Part 121 certificate holders have been provided an avenue through the SMSVP. The SMSVP standard, when properly applied, is the basis for formal State recognition of a certificate holder's SMS. While resembling Part 5, the SMSVP standard is a separate document used by the Flight Standards Service (AFS) SMS Program Office (SMSPO) to evaluate SMSVP participants.

The **SMSVP standard** details the minimum conformance expectations participants must maintain for State recognition of its SMS. Adherence to the SMSVP standard does not replace compliance with other FAA regulatory requirements. The certificate holder may also establish more stringent requirements in its system than those in this standard.

Therefore, an SMS required by regulation or developed within a voluntary program corresponds to ICAO SMS requirements and will be accepted by other ICAO member states, as long as aviation service providers under Part 5 or in the SMSVP meet program requirements specified for FAA recognition. For those aviation service providers who are adhering to the SMSVP, the FAA has created a guide that allows FAA recognition. This guide is called the **SMSVP Guide**. The importance of having a path for non-Part 121 operators to meet ICAO SMS requirements must not be understated. Instead of regulation, industry has promoted and encouraged the use of SMS principles and recognition. Flight training organization are being asked by potential foreign customers where they are at within their SMS, flight operations that contract with the federal government are requiring SMS recognition in order to continue to do business together, corporate flight departments who operate internationally are expected by the foreign countries they visit to meet SMS requirements, and aviation related design and manufacturer companies are recognizing the need to understand and implement SMS principles. As FAA inspectors are trained and become more familiar with SMS principles, the questions they ask, such as, "why didn't your safety management activities identify this problem and if it was identified, why did the management system not contain and/or correct this problem?". This alone will promote an understanding and eventual use of SMS principles, regardless of whether or not it is required by regulation.

For the purposes of this chapter, a majority of the time will be spent on exploring the process outlined under the SMSVP Guide to proceed to **active conformance**. The final level in which your organization's SMS is acceptable to the state. We will also introduce and compare Part 5 requirements to reach compliance, but will not delve into the details beyond AC 120-92, because airlines have already begun this process with the compliance deadline of March 9, 2018, quickly approaching; the FAA has also provided extensive guidance under Part 5, 8900.1 Volume 10, "Safety Assurance System Policy and Procedures," and 8900.1 Volume 17, "Safety Management System."

SMS VOLUNTARY PROGRAM

Many operators within the United States began their SMS journey under the SMS Pilot Program and the SMSVP. Many airlines that initiated their SMS under the SMSVP found the transition to Part 5 much easier, because the SMSVP was just that, voluntary. It actually gave the FAA more flexibility and guidance to the Part 121 Air Carriers. As Part 5 was released, it resulted in law being established, with written law being established the guidance, methods, and expectations had to mirror Part 5. As it made it easier for those who took the initiative to develop their SMS under the SMSVP, it also has made it harder for new Part 121 to get up to standards and for the FAA to provide guidance that does not overstep their bounds established within the law.

This process to reach active conformance is not a day trip but a long journey, a marathon. It requires clear commitment and fortitude from the leaders within the organization. They need to continue forward to consistently improve and update the entire organization, all the while developing, implementing, and maintaining system safety principles and a safety management system. With this said, remember, an SMS does not have to be large, complex, or expensive in order to add value. If there are active involvement of the operational leaders, open lines of communication up and down the organization and among peers, vigilance in looking for new operations, and assurance that employees know that safety is an essential part of their job performance, the organization will have an effective SMS that helps decision makers at all levels.

General Information

As you get started, new terms, that represent groups of people or entire organizations, will need to be understood. These terms include:

- **Certificate Management Team (CMT).** The CMT is responsible for validating the certificate holder's management system applications during both the implementation process and after full implementation. Office management is responsible for allocating the resources to accomplish this requirement.

  This consists of your local FAA personnel at the FAA FSDO in your state. The FSDO is sometimes referred as the certificate management office/unit (CMO/CMU) or certificate-holding district office (CHDO). Your FAA FSDO will appoint, key personnel to oversee and approve the process, this is your CMT. The FSDO will also select a key individual, within the CMT, who has the greatest understanding and commitment to supporting your organization. This individual will be named the point-of-contact (POC) to oversee CMT validation activities and communicate with the SMS Program Office (SMSPO) at a national level.

- **SMS Regional Office Point of Contact (RPOC).** Your local FSDO may not be trained or familiar with SMS processes for implementation, therefore the FAA has an SMS RPCO as well. The RPOC is the primary resource for SMS implementation and support above your local FSDO. RPOCs are alerted to requests for SMSPO onsite assistance by your organization or the CMT and may participate in those activities. RPOCs may be asked to address and help resolve SMS related conflicts so they do not have to be addressed on the national level through the SMSPO. Additionally, RPOCs are information resources on SMS trends, development, and news throughout their region, such as, their quarterly SMS newsletter available online.

- **SMS Program Office (SMSPO).** The SMSPO is the final authority on application of the SMSVP Standard and at the last stages of gaining SMS recognition from the FAA, the SMSPO will conduct the final validations to ensure you meet SMSVP requirements. This validation will require a joint effort from the CMT and the SMSPO resulting in a 2-3 day FAA onsite visit. The SMSPO may be contacted for guidance and policy interpretation through the CMT. SMSPO support is readily available upon request for all pre-application, validation, and continued operational safety (COS) activities.

  The SMSPO is part of the AFS-900 National Field Office. FAA assigns the SMSPO as the office of primary responsibility and focal point for AFS-900 SMS initiatives including: certificate holder implementation; SMS policy and guidance development; and SMS rulemaking. The SMSPO issues advisory and outreach materials to assist FAA field organizations and aviation service providers in SMS development and implementation. 14 CFR Part 5 is the SMSVP standard foundation document.

- **Certificate holder.** A certificate holder is authorized by the FAA to provide an aviation service or product. This is a FAA term, to fit within FAA terminology. Certificate holders are those organizations that have some type of certificate administered by the FAA. A Part 141 pilot school has an air agency certificate issued by the FAA to operate under Part 141 of the Federal Aviation Regulations. A Part 135, 145, and others have similar certificates. The FAA categorizes whom they are working with based on the FAA certificates they hold. This term does not lend itself to the idea of the System. For example, an aviation university may hold multiple FAA certificates, Part 141 and Part 145; and if they desired they could obtain a Part 135 Certificate as well. They may also operate under Part 61 and conduct remote pilot operations that have no certificate requirements to date. This can cause the term certificate holder to be confusing, because SMS clearly relates to the entire system (i.e. the entire organization) regardless if certain parts (i.e., 141, 145, 125, etc.) have FAA certification or acceptance attached. While a term such as "aviation service provider" may seem more appropriate, for consistency in this book the term certificate holder will continue to be used, while it is understood the SMS will affect all parts of the organization not just the Part 141 portion for example.

  In SMS development, a certificate holder designates an accountable executive who has final authority over operations authorized under its certificate/organization and is ultimately responsible for their company's safety performance. He/she signs and submits the SMS implementation plan on behalf of their company. The accountable executive's signature is a commitment to provide adequate resources for SMS development, implement SMS in all relevant areas of their organization, and ensure ongoing conformance to the SMSVP Standard.

- **Senior Technical Specialist (STS).** The senior technical specialist for safety management is the FAA's senior SMS subject matter expert and the official Aviation Safety Organization (AVS) SMS advisor. Generally speaking the STS will not be a main POC but another resource your CMT, RPOC, or even the SMSPO may go to. The STS resides in the AFS-900 National Field Office. The STS consults on all internal and external SMS development, technical issues, rulemaking, and policy formation. In addition to the SMSPO, the STS works closely with industry, government agencies, advocacy groups, and international organizations to advance SMS and its application within the National Airspace System.

A certificate holder may develop and implement an SMS in any manner it deems appropriate. When a certificate holder requests FAA recognition of its SMS, however, an implementation plan is submitted to its CMT for validation against the SMSVP standard

Once started, the certificate holder is expected to make steady progress towards full SMS implementation and continual improvement. Generally speaking, plan for no more than 12 calendar months between each category. The following categories denote the progress expected:

- **SMSVP Active Applicant.** The certificate holder and CMT have committed to sufficiently support the SMS implementation and validation processes.
- **SMSVP Active Participant.** The certificate holder officially begins and maintains its implementation efforts.
- **SMSVP Active Conformance.** The CMT and SMSPO acknowledge full implementation of the certificate holder's SMS. The certificate holder is expected to use and continually improve its safety management processes.

The SMSPO has sole authority to authorize or withdraw recognition of a certificate holder's SMS. The SMSPO's primary objective is to assist CMTs in validating SMS development and certificate holders maintain their active conformance status. The SMSPO will maintain an SMSVP Status Roster of all active participants. When a certificate holder fails to meet SMSVP standards, it becomes an **SMSVP non-active participant**.

After SMS full implementation is recognized, the certificate holder is expected to use and continually improve its safety management processes. The CMT is expected to perform its certificate monitoring duties where SMS is one of a number of performance measures determining **continued operational safety (COS)**. The SMSPO periodically verifies the certificate holder's conformance to the SMSVP Standard by oversight data collection.

SMSVP participants are free to withdraw at any time. If the certificate holder withdraws after SMSVP acceptance, it must notify their CMT and the SMSPO and their status will be changed to "voluntary withdrawal" and the effective date recorded in the status roster.

SMSVP implementation uses a phased approach. The following process steps are detailed throughout SMSVP document. We will be conducting a summary of the following phases:

1. Preparation phase.
2. CMT validation phase.
3. Documentation validation phase.
4. Performance demonstration phase.
5. Administrative process phase.
6. COS phase.

Preparation Phase

As with most important things, the time taken to prepare will provide a return exponential to the time it takes to set the stage. As you begin to establish an SMS at your organization, start setting the groundwork. Having an SMS is more than being able to put out a news release, gaining recognition, or even obtaining more opportunities for business. *Our incentive must be that you believe, by implementing and maintaining an SMS, that you will ensure greater quality in the service or product you provide and you believe it will make your organization better.* If you are doing this because you have to, the journey will start feeling more like an iron man decathlon instead of a marathon. Your vision regarding where you are heading is more than active conformance, and ultimately you will be able to say, "This is just how we do business."

In the context of the four components of SMS, consider safety promotion. What are the views of leadership? Are they ready to support the effort needed to start the SMS marathon? Are they communicating the willingness to provide the resources necessary to implement and maintain SMS? Are they willing to establish a system-wide safety policy regarding SMS and will they provide the resources to train key personnel to begin the process?

Once you have clearly established organizational buy-in from the accountable executive within the organization, your next area of clear communication must start with your local FAA FSDO. The FSDO will be a key partner in reaching active conformance. The FSDO will help you develop your CMT; all efforts will be coordinated through them. Take the time to meet with your FSDO, discuss your plan, and identify their interest, knowledge of SMS and ability to support and provide the resources necessary to begin the process. Your FSDO may still be learning about SMS, so invite them to view your organization's training sessions to help them better understand how your system operates. If they don't have the expertise or resources to assist, work through them to the RPOC as well as the SMSPO to provide additional guidance. Keep the two-way line of communication open with the FSDO, regardless of whether you are talking to them, the regional office, or the national office. Remember when you reach active conformance, it will be the CMT that will provide the continuous oversight to maintain your active conformance status.

Once you have the buy-in from both the organization accountable executive (i.e., the certificate holder) as well as the CMT, indicate to the CMT your intention to implement SMS. The CMT will then communicate with the SMSPO national coordinator to begin the application process. The national coordinator will ensure all relevant parties are informed of the certificate holder's requested entry into the SMSVP. At this point, you will know how well you communicated your intentions not only with your organization but also your CMT.

In order to continue the application process, the organization (namely the certificate holder) and CMT must commit, in writing, to providing sufficient resources to ensure successful SMS implementation. The SMSPO will provide both information describing the SMSVP validation process and respective roles, responsibilities, and expectations as you move forward.

Once the certificate holder and CMT completely review SMSVP information they must commit to supporting the SMS implementation process. Without a firm commitment from both parties, SMSPO communications will be limited to promotional materials. A letter or e-mail from CMT management and the certificate holder's executive management to the SMSPO National Coordinator is considered a documented commitment. When the respective CMT and certificate holder commitments are received, the SMSPO will designate the certificate holder as an SMSVP active applicant. This permits AFS offices to expend resources supporting the SMS implementation plan and CMT validation planning support. Unfortunately, this is the only resource it provides. The local FSDO may not have the human resources to train their employees in SMS for the voluntary program, or the time to perform their duties as well. You may have to slow down your timeline and utilize resources on a regional (RPOC) or national level (SMSPO).

Upon reaching active applicant status, the SMSPO national coordinator will identify an SMS implementation support team (IST) to conduct an initial workshop with the certificate holder and CMT. Before the workshop, the IST will provide copies of all applicable documents and information it expects you to reference. This material can seem overwhelming; the FAA has provided mountains of documentation to assist organizations, yet without significant expertise in SMS it can seem redundant and quite daunting. The workshop is the opportunity to get your questions answered. SMS requires continual buy-in, so include those who need to know and those you hope to help you lead the effort to accomplish the SMS implementation.

When the initial workshop is scheduled your organization's point of contact will receive an email stating:

1. Commitment dates;
2. Location;
3. Service providers name and address (organization name/certificate holder);
4. Service Provider's POC;
5. CMO, which is your FSDO;
6. CMO POC;
7. Certificate holders principle inspectors (who know your organization the best);
8. Notes (what the purpose of visit is); and
9. IST members from the regional as well as the national SMS program office.

The IST will conduct a multi-day workshop. Part of the workshop is just for the CMT to address "FAA Only" issues. The remaining workshop time is for the certificate holder and CMT to address the following:

1. Organizational concepts and considerations;
2. Description of the SMSVP standard;
3. Description of service provider SMS tools;
4. The SMSVP implementation and validation processes;
5. SMS Active Participant Acknowledgement; and
6. COS oversight expectations.

At the completion of the multi-day workshop, the certificate holder should be clearly aware of the work required to move forward. As previously discussed, the amount of work will depend largely on the complexity of the organization and the internal safety processes already in operation. For example, many organizations already have a safety reporting available to employees. That portion of the SMS will already be complete. The key to the SMS implementation plan is that **it must be developed in a form, manner, and medium that meets its needs and is agreed to by the CMT.**

The certificate holder's implementation plan is a "roadmap" describing actions needed to conform to the SMSVP Standard. The implementation plan should detail a realistic timeline. The certificate holder should examine its organizational structure and manuals to identify individuals responsible for process designs and have authority and technical expertise to apply those designs.

It is incumbent upon the certificate holder to identify individuals responsible for developing, implementing, and maintaining SMS processes within their respective areas of responsibility. Process owner and manager responsibilities include:

1. Hazard identification;
2. Safety risk assessment, and risk acceptance;
3. Evaluating the effectiveness of safety risk controls;
4. Promoting safety; and
5. Submitting performance reports to the accountable executive on SMS performance.

Implementation plan outlines should include:

1. A listing of the relevant sections of the SMSVP standard and associated reference sources;
2. A brief narrative describing where processes conform to the SMSVP standard, or what actions the certificate holder will take to comply;
3. Identification of specific employees that will be responsible for implementing required actions; and
4. Estimated target dates that each expectation will be ready for design validation and performance demonstration.

The certificate holder's implementation plan is the result of a thorough system wide gap analysis. A gap analysis compares existing processes, procedures, programs, and activities to the SMSVP Standard. The gap analysis simplifies development of an effective implementation plan. Completing a gap analysis allows the certificate holder to determine what existing programs, processes, and practices comply with the SMSVP standard and identify those that do not.

Gap Analysis

Your gap analysis tool(s) will be critical to your success. A current version of the FAA gap analysis for Part 121 air carriers and maintenance repair organizations can be found in the Reader Resources for this book. This tool is Microsoft Excel based with multiple tabs to help you identify the "gaps" in your organization in relation to meeting SMS active conformance. Initially, it can feel overwhelming to use this tool, but it is a necessity. Remember the importance of buy-in as well. Don't assign a work-study to do this, as this is a joint project that must involve members from all departments of your organization.

You will note the gap analysis tool systematically asks questions about your organization, of which you must decide by the grading criteria how close you are to meeting the active conformance criteria. The distance you are from active conformance is your "gap." Once you have identified the gap, you will need a clear implementation plan regarding how you are going to meet the standard.

The gap analysis tool has five distinct parts that should be addressed chronologically:

GENERAL INFORMATION/SYSTEM SEGMENT IDENTIFICATION. Before beginning the preliminary gap analysis, the organization should accurately identify its system segments as they pertain to the scope of the SMS. The system segments are the columns within your organization; it is the one part of the gap tool that you can change to align with your system. This should be a true reflection of the different parts of your organization. For example, a university flight school might have the following system segments:

1. Flight operations;
2. Maintenance operations;
3. Academics; and
4. Safety.

These segments are completely tailored to your organization and each organization will be unique. The key is making sure it accurately reflects your organization. More columns means more up front work, but the final product will be better if it is comprehensive. The gap analysis Tool will use these system segments throughout the analysis to identify gaps between the organization's current state and the design and manufacturing (D&M) SMS framework elements and sub-elements.

When conducting any SMS gap analysis it is critical that you understand the four components of SMS. The gap analysis questions will seem strangely similar, but you must not interpret the questions in a vacuum, rather recognize that each component plays a critical role in every procedure and process. Questions asked through the eyes of one component, may look entirely different than through the eyes of another component.

PRELIMINARY GAP ANALYSIS. The preliminary gap analysis tool assists the organization in conducting an initial high-level assessment on existing organizational programs, systems, and activities with respect to the components, elements, and sub-elements found in the functional expectations of the SMS. This initial assessment provides a first look in assessing compliance with the SMS. This may be started at your initial SMS workshop, where the SMSPO, RPOC, and CMO can assist you in the process. The organization and FAA representatives will identify gaps between the organization's current processes and compliance to the SMS framework. Specifically, the organization will provide its best guess assessment of each component, element and sub-Process in the preliminary gap analysis Tool. The assessment provides a starting place for the organization to conduct a detailed evaluation of the gaps (see detailed gap analysis) between current safety policies and the D&M SMS framework.

DETAILED GAP ANALYSIS. The detailed gap analysis Tool assists the organization with identifying gaps between the current processes and compliance with the D&M SMS framework. Similar to the preliminary gap analysis, the organization will perform the evaluation for all system segments, but the detailed gap analysis identifies specific Expectations within each component, element and sub-element within the SMS.

During the detailed gap analysis the organization will provide objective evidence for each Expectation rated "P" or above on the maturity status scale provided. Objective evidence includes physical or electronic documents, manual references, training material, records, correspondence (email, memo, etc.), organizational charts, and meeting minutes.

The detailed gap analysis includes expectations that represent what the FAA believes to be the basis for an SMS. In addition, the FAA has provided other information that can be considered to be developmental guidance. The FAA requests that participants identify a status and provide feedback on the both the expectations and guidance items. Feedback will be used to support future SMS rulemaking, policy, guidance material development, and oversight process, as applicable.

DETAILED EXECUTIVE SUMMARY. The detailed executive summary compiles the status identified within the detailed gap analysis and displays summarized information regarding SMS status. This tool can display a variety of summary options, including a high-level status of each component and a detailed summary of the elements and sub-elements

Implementation Plan

THE IMPLEMENTATION PLAN TOOL. This tool assists the organization in identifying tasks and activities that will address the gaps identified during the detailed gap analysis. While not required, the tool is intended to allow the organization to ensure all expectations that are identified within the D&M SMS framework are addressed when working on gap closures. The organization may find it useful to also develop a narrative version of a plan that outlines their overall approach to implementing SMS. Additionally, the organization may choose to add to this worksheet and/or add another worksheet that supports the tracking and completion of tasks/activities used to close the gaps.

IMPLEMENTATION ASSESSMENT. The implementation assessment tool compiles the results of the detailed gap analysis worksheet and provides additional fields for FAA observational notes and assessment. The FAA will use the organization's detailed gap analysis as a starting point for conducting the implementation assessment and draft the final implementation plan. The implementation plan will describe how the organization intends to close the existing gaps by meeting the objectives and expectations of the SMS framework.

IMPLEMENTATION PLAN SUBMISSION. Once the certificate holder has developed its implementation plan, it will be submitted to the certificate holding office for review. The time between the initial SMS workshop and the certificate holder's implementation plan can take as long as a year but may be completed sooner. Once the plan is agreeable to the CMT, and the SMSPO has completed its quality review, the participants are ready to start the validation phase.

CMT Validation Phase

The CMT will perform a review of the certificate holder's implementation plan using these general guidelines listed in the SMSVP Guide.

1. Is the system properly identified?
2. Does the plan result in meeting the SMS standards for active conformance?
3. Does the documentation match what is indicated?
4. Are the processes in place or to be used appropriate to the complexity of the organization?
5. Are there valid POCs for validation activities?
6. Can the CMT develop a viable **validation project plan (VPP)** based on implantation dates forecasted. The target date on the certificate holder's implementation plan is not necessarily the date that the CMT can perform validation work, but helps it forecast dates for validation activities.

CMT VPP

The objective of a good validation plan is forecasting the resources needed to perform appropriate validation activities on the certificate holder's safety management processes. To those ends, during implementation plan review the CMT will consider how to manage its validation work.

VPP development is based on the certificate holder's Implementation Plan submission. The submitted plan guides the CMT to identify its corresponding subject matter resources to validate specific areas of the certificate holder's management system. Based on the CMT manager's determination of available resources, the CMT POC can prepare a resource document for management review and approval. The resource document identifies the CMT and certificate holder's implementation plan POCs. Once appropriate resources are identified, the CMT can draft its VPP. (The CMT manager may adjust office resources to address VPP requirements.)

To develop a viable VPP, the validation team must understand the purpose and use of the design and performance job aids provided in this document. Validation and demonstration activities must be accomplished in sufficient detail to ensure conformance with the SMSVP standard. It is recommended the CMT contact the SMSPO for assistance in how to use the supplied design and performance job aids. The SMSPO can assist the CMT in determining effective and efficient ways to manage its validation work.

VPP Considerations

The CMT and certificate holder must agree on the VPP schedule of events. The CMT should design its validation activities to allow for assessment-correction-reassessment, as planned CMT design and performance validations dates may become unreliable if the proposed implementation plan timelines are not being met. The CMT and certificate holder POCs should collaborate throughout the validation process and revise the VPP schedule of events as necessary. Depending on the proposed validation activity there may not be a need to conduct independent validations for processes uniformly applied throughout the organization. These processes may require only a onetime sampling to validate an entire system application. Also, some validation work can be accomplished remotely, while other work requires site visits.

The SMSVP phased approach requires, that in collaboration with the SMSPO, two performance demonstration activities be scheduled at the very end of the CMT validation process:

1. The management review performance demonstration; and
2. The corporate SRM performance demonstration.

These final two validations are performed at the end of all validations with the SMSPO, CMT, and the certificate holder. Successful completion of these two validations, without corrections, result in the FAA recognizing the certificate holder's active conformance.

CMT and Certificate Holder Validation Planning Meeting

The CMT POC will organize an SMS validation-planning meeting with the certificate holder to agree on the proposed VPP timelines. Appropriate CMT and certificate holder implementation teams must attend to agree or revise the validation plan schedule.

1. The CMT POC will present its SMS VPP to the certificate holder and discuss planned activities. Certificate holder and CMT agreement of the SMS VPP represents mutual acceptance of the plan's timeline for completion of CMT validation activities. The certificate holder should also commit to aggressively work toward implementation plan completion.

2. The certificate holder and CMT should discuss how implementation plan changes might affect VPP activities.

3. The CMT will notify the certificate holder of the design validation and performance demonstration job aids being used to validate their SMS, and how they should be used. The CMT will remind the certificate holder that they must provide evidence of its own internal assessments before the CMT validates those processes.

4. During the validation-planning meeting CMT and certificate holder concerns are addressed. Both will agree that all the planning requirements are complete and the certificate holder is ready to be acknowledged as an active participant by the SMSPO. The CMT will forward its acknowledgement recommendation to the SMSPO, the certificate holder's SMS implementation plan, and the CMT's VPP.

SMSPO VPP Review

Once the SMSPO has received the certificate holder's implementation plan, the CMT's VPP, and CMT acknowledgement recommendation, the SMSPO performs a quality review. The SMSPO will contact the CMT POC if there are any questions or open issues from its review. Any subsequent corrections of deficiencies will be coordinated with the impacted parties, as applicable.

When the SMSPO determines the certificate holder has a complete and approvable implementation plan, it will issue a letter acknowledging the certificate holder as an **SMSVP active participant** and update the SMSVP status roster.

Documentation Validation Phase

The documentation validation phase is to determine if the certificate holder has an adequately designed SMS with the required safety management activities and processes in their organizational system. As the word implies, design focuses on what is documented and is not as concerned with whether or not it is actually performed. After looking at what is designed, the next phase, deemed the performance demonstration phase, will identify if the design is functioning or performing as expected.

The design job aids in the SMSVP will be used to evaluate certificate holder's documentation describing its SMS applications. Even though design validations may occur at different times, and on different certificate holder process areas/departments, they cannot be considered complete until there are enough validation records to demonstrate conformance with the SMSVP standard.

The CMT will validate, to the extent possible, the certificate holder's process design conforms to the SMSVP standard. The design job aids encompass the operational SMSVP conformance requirements. These job aids are considered the minimum performance validation activities to be used during the design validation phase.

SMS Design Job Aid—Frequency of Use

The number of design job aids to be completed will be identified in the CMT VPP. The following guidelines can be expected from the CMT and when your organization conducts its own assessment this should be considered:

1. While some job aids may be completed just once, others may be completed multiple times for multiple process areas/departments.
2. Certificate holder processes that generally apply across the entire organization require only one design validation.
3. Certificate holder processes that apply to specific process areas/departments require design assessment for each process area/department (process area SRM, process area continuous monitoring.).
4. When the design job aid questions are all answered affirmatively for a process area(s), the CMT can prepare for its performance demonstration on that process area(s).

Performance Demonstration Phase

The performance demonstration phase is to determine whether the certificate holder's process applications have been applied operationally and are working as designed. While certificate holders are not required to do these job aids, they are required to complete their own internal performance assessments before CMT performance validation assessments begin. Evidence of these certificate holder internal performance assessments must be made available to the CMT upon request.

The performance job aid will be used to evaluate the certificate holder's safety management performance. Where actual field performance cannot be assessed (emergency response plans), the CMT is permitted to use simulated processes (sometimes called "table top exercises") allowing CMT to evaluate the certificate holder's capabilities without an actual performance demonstration. The performance job aids provide the inspector sample validation activities to assess certificate holder conformance with the SMSVP standard.

Even though performance demonstrations may occur at different times and on different certificate holder process areas/departments, they cannot be considered complete until there are enough observations to demonstrate system-wide conformance to the SMSVP standard.

The CMT will validate, to the extent possible, that the certificate holder's process applications actually function in day-to-day operations. The CMT has been provided with performance job aids covering key SMSVP operational compliance requirements. While these job aids may not completely evaluate every SMSVP requirement, they are the minimum performance validation activities to be used during the performance demonstration phase.

The number of performance validations to be completed will be identified on the CMT VPP. That said, the CMT might add performance validations at its discretion. While some validation activities may be completed one time, others may be completed multiple times for multiple process areas/departments (such as policy work, safety policy, emergency response plan).

As you work through your job aids, you will see that safety policy generally applies across the entire organization and should require only one performance demonstration.

When the certificate holder receives a "satisfactory" evaluation on the all of the assigned performance tests the CMT can acknowledge the certificate holder's SMS performance capability.

Combined CMT and SMSPO Performance Validations

The stated earlier, the CMT may independently accomplish all performance demonstrations except for two that must be completed in collaboration with the SMSPO:

1. Management review process performance demonstration; and
2. Corporate SRM process performance demonstration.

These demonstrations are completed as a "table top exercise" with appropriate representatives of the certificate holder and CMT, and SMSPO IST participating. The certificate holder's accountable executive must participate (in person or virtually) for the management review demonstration. These are the final performance demonstrations that must be completed to close out the CMT's VPP.

These two performance demonstration exercises are critical and all the work to create buy-in from the accountable executive and other leadership will pay off. If you have been communicating, training, and creating a culture to support SMS, it will be evident at this point. Furthermore, give leadership significant warning of what will occur. Schedules must be aligned; the accountable executive must be present. Communicating leadership's final role in implementation will naturally cause them to be more attentive through the beginning phases.

When the FAA SMSPO participates, they will not be asking the SMS experts, they will focus their conversation on the accountable executive and other key leaders (process owners and managers) within the organization that are defined in the safety policy. It can't be faked. They will be responsible to effectively conduct a management review and a corporate SRM as detailed in the job aids located in the SMSVP.

Administrative Process

Once all SMS design and performance validation activities are successfully completed, the CMT will close out the validation process by completing the following actions:

1. The CMT POC archives the final CMT VPP with attached certificate holder implementation plan following CMT local office policy and sending a copy to the SMSPO.
2. CMT management will recommend the SMSPO issue final recognition of the certificate holder's SMS.

Once the SMSPO receives the final VPP/implementation plan documents and the CMT manager's request for certificate holder recognition as having a fully implemented SMS, it will complete its administrative process review. This ensures that all SMSVP required administrative tasks have been completed. Any issues or required corrections will be coordinated with the CMT POC.

Upon satisfactory review, The SMSPO will change the certificate holder's status from SMS-VP active participant to SMSVP active conformance and issue the certificate holder a current status letter. (See Figure 7-1.)

The SMSPO will post a record of the certificate holder's SMSVP Active Compliance Status on the internal FAA SharePoint site. This site is accessible to all FAA personnel. The certificate holder's status is posted and constitutes evidence of "state recognition" of their SMS. This information may be provided to ICAO member states, if requested.

U.S. Department
of Transportation
**Federal Aviation
Administration**

March 9, 2016

Director of Safety
Flight School
7005 132nd Place SE
Newcastle, WA 98059

Dear Director of Safety,

This letter acknowledges Flight School as having an accepted Safety Management System (SMS) recognized by the Federal Aviation Administration (FAA) in accordance with the requirements set forth in the SMS Voluntary Program.

Formal SMS development consists of voluntary SMS implementation by operators and other aviation service providers using FAA stated standards. Those standards are based on FAA Order VS 8000.367A, Aviation Safety (AVS) Safety Management System Requirements, the SMS framework specified by the International Civil Aviation Organization (ICAO) in ICAO Annex 19, and further detailed in ICAO document 9859, Safety Management Manual (SMM). Participation in this program signifies that Flight School is implementing an SMS aligned with international standards.

Based on our review of Flight School's planning, documentation, and activities, we have determined that your SMS implementation meets the expectations of the Flight Standards SMS Voluntary Program guidance for acknowledgement of a fully functional SMS. Your FAA certificate management team (CMT) and the Flight Standards SMS Program Office validated this achievement with your cooperation.

The FAA SMSPO and your CMT congratulate you on your company's significant accomplishment in implementing a fully functional SMS that is "accepted by the State" in accordance with international requirements. To maintain your company's current "Active Conformance" status, your company must continue to apply the SMS processes that you developed and implemented to meet the FAA's SMS Voluntary Program requirements. Thank you for your continued commitment to improve aviation safety in our National Airspace System and again, congratulations on your momentous acheivement.

Sincerely,

Manager, Flight Standards
National Field Office

Figure 7-1 FAA letter of acknowledgment

COS Phase

Now that active conformance has been reached, as an organization you must continue to make sure that the procedures and process are integrated into everyday business. Continuous improvement must be your goal. It is imperative to practice and consider all four SMS components. At this point, the CMT will be your only necessary POC to perform continued observation of the certificate holder's applied safety management processes.

Failure of either the certificate holder or the CMT to adequately meet its oversight obligations on the certificate holder's SMS may result in SMSPO withdrawal of state recognition.

This means, as an organization, you can't wait for the CMT. You must have a schedule and you must continue communication with the CMT to ensure you maintain your active conformance status.

As discussed, earlier safety management processes and activities will continue to be integrated into the certificate holder's technical processes. As this happens, the CMT must broaden the scope of its normal surveillance activities to include safety management assessment. The more inspectors practice the inclusion of safety management oversight into their day-to-day surveillance activities, the easier and more intuitive the whole process becomes.

CMT must record all safety management assessment activities to demonstrate certificate holder conformance with the SMSVP standard. CMT surveillance activities, associated with safety management, must be recorded in the applicable FAA data repository. The safety management COS job aids are provided in the SMSVP Guide to assist the CMT with these activities.

There are three types of safety management job aids with specific applications depending on the type of surveillance activity being conducted:

- **Design Validation Job Aids.** These job aids are the same as used for initial SMS design validation. While the primary use of the aid is to evaluate initial compliance of a certificate holder's SMS process design with the SMSVP standard, the CMT may decide to use the job aid when safety management processes are revised (e.g., change in SRM procedures, change to safety assurance processes, etc.). The design validation job aid will assist the CMT in determining that the certificate holder's safety management processes maintain conformance to the SMSVP standard.
- **Performance Job Aids.** These job aids are used for safety management activities having a "corporate wide impact" on the system. The records of these processes are usually centrally located within the organization (e.g., corrective/preventive actions, management review actions, process changes resulting from the investigations process, audits findings and correction, etc.).
- **Supplemental Process Performance Job Aid.** This job aid is used in conjunction with all other technical process inspection job aids. The questions are applicable to multiple employee levels (e.g., local management, line employee, etc.).

PART 5 FOR AIR CARRIERS

The majority of this chapter was concentrated on exploring the process outlined under the SMSVP Guide to proceed to active conformance, the final level at which your organization's SMS is acceptable to the State. With the compliance deadline of March 9, 2018, Part 121 carriers should have implementation plans submitted and processes to validate compliance near completion. Understanding that the SMSVP process will make it easy for an individual to transition into the requirements of Part 5.

The FAA provides detailed guidance on the certification process of Part 121 air carriers. During the certification process, the CHDO and AFS-900 team will form a CPT. The CPT will follow the CPD found in Volume 10, Chapter 6, Section 2. Under no circumstances will an applicant be certificated until the CHDO, the regional RFSD offices, and AFS-900 are confident that the prospective air carrier is able to provide service at the highest degree of safety in the public interest.

The certification process for CFR 14 Part 121 applicants consists of a pre-application process and a series of four phases and three gates that must be successfully completed when progressing between phases.

The certification process flow chart, Figure 7-2 provides an overview of the certification/application process. Simultaneous use of the flow chart and narrative discussion will assist you and the inspector in understanding the certification process. The chart is particularly useful in determining whether the schedule of events is reasonable in terms of sequence, timeliness, and inspector resource availability. It also provides a perspective on how a particular event affects other events and is an important reference for planning various activities during the certification/application process.

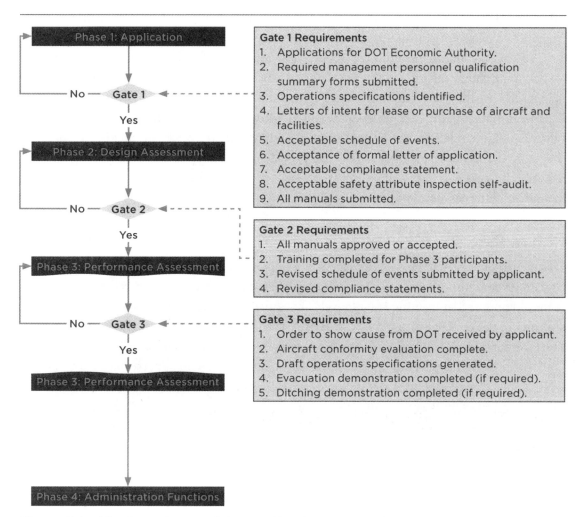

Figure 7-2 The four phases of the certification process for Part 121 air carriers.

Pre-Application Process

This part of the process is the most complex, demanding, and time consuming. Your objective during the pre-application process is to work with personnel at the CHDO to gather the required information and documentation necessary to prepare for the application process.

A vital part of this process is to gather information and gain an understanding of the importance of implementing an SMS. In addition, the applicant should review important voluntary safety programs such as the ASAP, the FOQA, and the voluntary disclosure reporting program (VDRP).

Initial Inquiries and Pre-Application

Initial inquiries about certification or requests for application may come in various formats from individuals or organizations. These inquiries may be in writing or in the form of meetings with CHDO personnel. Requests for applications may come from inexperienced and poorly prepared individuals, from well-prepared and financially sound organizations, or from individuals and organizations ranging between these extremes.

Upon initial contact, CHDO personnel should provide the applicant with a pre-application statement of intent (PASI) (FAA Form 8400-6) and instruct the applicant on how to complete the PASI. The CHDO personnel should also provide the applicant with the address of the certification process page and advise them that the information found in this website will assist them during the certification process.

Phase 1: Application

This phase begins when the applicant submits a request for a formal application meeting to the CHDO. During this phase, the AFS-900 team briefs the CHDO on the certification process. A formal application meeting is tentatively scheduled after the FAA receives all submissions required in the pre-application checklist (PAC). The CPT reviews the applicant's PAC submissions for completeness and accuracy before confirming the formal application meeting date. During the formal application meeting, the applicant's management personnel must demonstrate knowledge of their air carrier's system design. Phase 1 ends when the CPT accepts the formal application package and all Gate I requirements are met.

Phase 2: Design Assessment

The CPT evaluates the design of the applicant's operating systems to ensure their compliance with regulations and safety standards, including the obligation to provide service at the highest level of safety in the public interest. This phase uses safety attribute inspections (SAI) to collect data that will be used to determine if the air carrier's system design meets all regulatory requirements. Phase 2 ends when all manuals have been accepted or approved, and all Gate II requirements have been met. It should be noted, this process is fluid, the Part 121 operator and the FAA will be in constant communication, before the manuals have been officially submitted, there will have been significant communication conducted. When the Part 121 operator submits their manuals they are relatively confident they will be accepted or approved, with at most minor changes.

Phase 3: Performance Assessment

Aviation safety inspectors (ASI) use element performance inspections (EPI) during this phase to collect data that will be used to determine if the applicant's systems are performing as intended and producing the desired results. This phase requires the operation of an aircraft to aid in the assessment of the applicant's system design. Proving tests begin only after all Gate III requirements are met. Phase 3 ends after the successful completion of the proving tests.

Phase 4: Administrative Functions

The FAA issues an air carrier or operating certificate and operations specifications when you have completed all requirements, phases and gates of the certification process.

SCENARIOS FOR GROUP DISCUSSION

For the scenarios below, complete each statement. Imagine how continued oversight surveillance would look using SMS principles.

1. When an inspector(s) conducts a base inspection, they may determine what safety management processes or activities are assessed by....

2. When an inspector(s) conducts a flight check, they may ask site specific questions of employees such as... (Name three examples.)

3. CMT inspectors will soon realize that evaluating the certificate holder's ability to identify and correct its own problems before they are discovered by the FAA is of greater benefit to public safety than depending on the limited surveillance resources of the local CMO, consequently on the their next visit they...

4. The FAA is made aware of a loss of directional control incident on an uncontrolled airport. On their next visit to the certificate holders operation, what information might they ask for?

5. If an inspector finds a regulatory violation or process nonconformance, his or her most important concern is...

REVIEW QUESTIONS

1. Define the following key terms:
    a. Active applicant.
    b. Active conformance.
    c. Active participant.
    d. Gap analysis.
    e. Implementation plan.

2. Compare and contrast the design and the demonstration of the performance phases.

3. During which phase would one expect to receive the recognition of active applicant under the SMSVP?

4. What does it mean to be a non-active applicant?

5. In regard to the SMSVP, what two demonstration performance job aids must involve the SMSPO?

6. Who has final authority regarding recognition of an organization as reaching active conformance?

7. In reviewing the SMS gap tool, what segments or columns apply to your organization or to various organizations who might decide to obtain an SMS?

CHAPTER 8
Transitioning Your Safety Program to a Safety Management System

OBJECTIVES

- Explain the five levels of implementing an SMS.
- Explain the phased approach used in the SMSVP Guide.
- Explain the three reasons that justify using a phased approach when developing your SMS program.

INTRODUCTION

If your organization is developing an SMS, you probably have an existing safety program. This chapter will help your transition your safety program to an effective safety management system.

In the development and implementation of a SMS it is best to break down the overall complexity of the task into smaller, more manageable subcomponents. In this way, overwhelming and sometimes confusing complexity, and its underlying workload, may be turned into simpler and more transparent subsets of activities that only require minor increases in workloads and resources. This partial allocation of resources may be more commensurate with the requirements of each activity as well as the resources available to the service provider.

The reasons that justify a phased approach to SMS implementation can be expressed as:

- Providing a manageable series of steps to follow in implementing an SMS, including allocation of resources;
- Effectively managing the workload associated with SMS implementation; and
- Avoiding "cosmetic compliance."

> **Note:** The FAA's approach to a phased implementation of a SMS is based upon, but slightly different than the ICAO implementation strategy expressed in ICAO Document 9859, "Safety Management Manual (SMM)."

Your organization should set as its objective the realistic implementation of a comprehensive and effective SMS, not a token program. You simply cannot buy an SMS system or manual and expect the benefits of a fully implemented SMS.

According to the SMS Program Office, feedback from Pilot Project participants has shown that while full SMS implementation will certainly take longer, the robustness of the resulting SMS will be enhanced and early benefits realized as each implementation phase is completed. In this way, simpler safety management processes are established and benefits realized before moving on to processes of greater complexity. This is especially true with regard to SRM. In the reactive phase (level 2), a service provider will build an SRM system around known hazards which are already identified. This allows company resources to be focused on developing risk analysis, assessment and control processes (that frequently resolve old long term issues and hazards) unencumbered by the complexities necessary at the proactive (level 3) and predictive phases (level 4).

PHASED IMPLEMENTATION OF YOUR SMS

The following guidance for a phased implementation of SMS aims at:

- Providing a manageable series of steps to follow in implementing an SMS, including allocation of resources;
- Effectively managing the workload associated with SMS implementation;
- Pre-empting a "box checking" exercise; and
- Realization of safety management benefits and return on investment during an SMS implementation project.

Implementation Level 0: Orientation and Commitment

Level 0 is not so much a level as a status. It indicates that the service provider has not started formal SMS development or implementation and includes the time period between a service provider's first request for information from the FAA on SMS implementation and when the service provider's top management commits to implementing an SMS.

Level 0 is a time for the service provider to gather information, evaluate corporate goals and objectives and determine the viability of committing resources to an SMS implementation effort. Information requested from the FAA may be satisfied with emailed documents and/or reference material and/or referrals to Internet websites (www.faa.gov/about/initiatives/sms) where information/documents/tools may be downloaded.

Face-to face informational meetings between the individual service provider, responsible FAA CMT and the SMSPO are not normally conducted at level 0; however they may be conducted on a case-by-case basis depending upon FAA SMSPO resource availability and other circumstances.

In lieu of individual meetings, activities such as group outreach presentations and group seminars will be conducted in order to establish relationships and define SMS PP expectations for service provider's top management and oversight organizations.

Implementation Level 1: Planning and Organization

Level 1 begins when a service provider's top management commits to providing the resources necessary for full implementation of SMS through out the organization.

Gap Analysis Tool

The first step in developing an SMS is for the service provider to analyze its existing programs, systems, and activities with respect to the SMS functional expectations.

The gap analysis process should consider and encompass the entire organization (e.g., functions, processes, organizational departments, etc.) to be covered by the SMS. As a minimum, the gap analysis and SMS should cover all of the expectations of the SMS framework. The gap analysis should be continuously be updated as the service provider progresses through the SMS implementation process.

Implementation Plan

Once the gap analysis has been performed, an implementation plan is prepared. The implementation plan is simply a road map describing how the service provider intends to close the existing gaps by meeting the objectives and expectations in the SMS framework.

While no actual development activities are expected during Level 1 beyond those listed in the SMS Framework, your organization will organize resources, assign responsibilities, set schedules and define objectives necessary to address all gaps identified.

It should be noted that at each level of implementation, top management's approval of the implementation plan must include allocation of necessary resources.

At the completion of Level 1, your organization should have created and implemented the following documents:

- Management commitment letter;
- Safety policy documents;
- Comprehensive SMS implementation plan (summary) for the entire organization through SMS Implementation level 4; and
- SMS training plan for all employees.

Implementation Level 2: Reactive Process, Basic Risk Management

At level 2, your organization develops and implements a basic SRM process and plan, organize and prepare the organization for further SMS development. Information acquisition, processing, and analysis functions are implemented and a tracking system for risk control and corrective actions are established. At this phase, your organization corrects known deficiencies in safety management practices and operational processes and develops an awareness of hazards and responds with appropriate systematic application of preventative or corrective actions. This allows your organization to react to unwanted events and problems as they occur and develop appropriate remedial action. For this reason, this level is termed "reactive." While this is not the final objective of an SMS, it is an important step in the evolution of safety management capabilities.

At the completion of level 2, your organization should have created and implemented the following documents:

- Objective evidence that SRM processes and procedures have been applied to at least one existing hazard and that the mitigation process has been initiated;
- Updated comprehensive SMS implementation plan (summary document) for all elements to take the organization through level 4; and
- Updated SMS training plan document for all employees.

Implementation Level 3: Proactive Processes, Looking Ahead (Fully Functioning SMS)

Level 3 expects SRM to be applied to initial design of systems, processes, organizations, and products, development of operational procedures, and planned changes to operational processes. The activities involved in the SRM process involve careful analysis of systems and tasks involved; identification of potential hazards in these functions, and development of risk controls. The risk management process developed at level 2 is used to analyze, document, and track these activities. Because the service provider is now using the processes to look ahead, this level is termed "proactive." At this level, however, these proactive processes have been implemented but their performance has not yet been proven.

At the completion of level 3, your organization should have created and implemented the following documents:

- Objective evidence that SRM processes and procedures have been applied to all safety risk management operating processes;
- Objective evidence that SRM processes and procedures have been applied to at least one existing hazard and that the mitigation process has been initiated;

- Updated comprehensive SMS implementation plan (summary document) for all elements; and
- Updated SMS training plan for all employees.

Implementation Level 4: Continuous Improvement, Continued Assurance

The final level of SMS maturity is the continuous improvement level. Processes have been in place and their performance and effectiveness have been verified. The complete SA process, including continuous monitoring and the remaining features of the other SRM and SA processes are functioning. A major objective of a successful SMS is to attain and maintain this continuous improvement status for the life of the organization.

SMS Framework Expectation	Level 1	Level 2	Level 3	Level 4
Component 1: Safety Policy and Objectives		X		
Safety Policy	X			X
Management Commitment and Safety Accountabilities		X		X
Key Safety Personnel	X			X
Emergency Preparedness and Response		X		X
SMS Documentation and Records		X		X
Component 2: Safety Risk Management			X	X
Hazard Identification and Analysis		X		X
Risk Assessments and Controls		X		X
Identify Hazards		X		X
Analyze Safety Risks		X		X
Assess Safety Risk		X		X
Control/Mitigate Safety Risk		X		X
Component 3: Safety Assurance			X	X
Continuous Monitoring		X		X
Internal Audits by Operational Departments		X		X
Internal Evaluation		X		X
External Auditing of the SMS		X		X
Investigation				X
Employee Reporting and Feedback System		X		X
Analysis of Data		X		X
Preventive/Corrective Actions		X		X
Management Review		X		X
Continuous Improvement		X		X
Component 4: Safety Promotion			X	X
Personnel Expectations (Competence)			X	X
Training		X		X
Communications and Awareness		X		X
Establishment of a Culture of Safety				X

Table 8-1 SMS implementation.

REVIEW QUESTIONS

1. Explain the three reasons that justify using a phased approach when developing your SMS Program.

2. Explain the five levels of implementing an SMS.

3. Explain the phased approach used in the SMSVP Guide.

CHAPTER 9
Developing a Safety Policy for Your Organization

OBJECTIVES

- Identify the elements in an effective safety policy statement.
- Describe how to write safety policy that meets FAA requirements.
- Explain the significance of safety objectives and how they should be developed to complement your safety policy.

KEY TERMS

- Designated Management Personnel
- Emergency Response Plan (ERP)
- Process Manager
- Process Owner
- Safety Performance

INTRODUCTION

One of the most difficult aspects of developing safety policies, a safety policy statement, and safety objectives is getting consistent involvement from leadership. In general, the idea of developing and writing policy, establishing a safety policy statement, and safety objectives are foreign to most accountable executives. It can seem menial and even a sort of technicality, when in fact it lays the foundation for the rest of your safety management system. In establishing safety policy and especially creating safety objectives, "starting with the end in sight" is sound advice. If the beginning sets the stage for the end, it further supports the need for the accountable executive and key leadership to be involved.

Within the following chapter we will break down the difference between a safety policy and a policy statement, identifying minimum requirements for documentation, establish how to develop safety objectives and emphasize the importance safety objectives have in maintaining your SMS.

THE SAFETY POLICY STATEMENT

The safety policy statement is a concise document from the accountable executive that conveys the organization's basic commitments to safety management. It provides a basis for more detailed setting of objectives for planning and performance measurement, assignment of responsibility, and reporting, including clear statements regarding behavioral and performance expectations.

The safety policy statement is part of the overall safety policy. The safety policy statement will be supported by additional safety policy that expands in specifics areas, and, where applicable, it may also set out procedures. (An example of a safety policy statement can be found later in this chapter.)

14 CFR Part 5 and the SMSVP establish the minimum content that comprises the safety policy statement. This policy statement must specifically state the entire expectations outline in Part 5 and the SMSVP. A common mistake is to consider safety policy that is written outside the specific safety policy statement to be acceptable by the FAA. It is not. The FAA desires to see all components required by Part 5 or the SMSVP clearly contained within the safety policy statement and signed by the accountable executive.

This is one area where the natural tendency of the FAA and pilots in general is to reduce SMS to checking off the required boxes of their checklists to validate an effective SMS safety policy. An SMS is exactly as the title states, a system, yet the FAA in our experience is adamant that every safety policy statement must be almost verbatim in its structure, including the required items and ending with a commitment signature from the accountable executive. It is understandable to have this expectation, but the organization must recognize this simply defines the framework of the overall safety policy for the organization.

Listed below are the minimum requirements identified in §5.21 and SMSVP as well as an explanation of each. The certificate holder must have a safety policy statement that includes at least the following:

- **The safety objectives of the certificate holder.** This step must come early in the SMS process. Certificate holders/service providers must develop safety objectives as part of the actual safety policy statement. The assessment process required by §5.73 requires clear safety objective so that decisions are made regarding attainment of these objectives. If you don't have objectives regarding where you are heading, as you begin manage of safe you will have no significant method to measure that safety.

- **A commitment of the certificate holder to fulfill the organization's safety objectives.** As in the following paragraph, this shows the top-down management approach, recognizing if you do not have commitment from the accountable executive, the likelihood of your SMS working is minimum. This holds the accountable executive accountable to meet and support the SMS to meet the safety objectives.

- **A clear statement about the provision of the necessary resources for the implementation of the SMS.** A resource commitment within the safety policy that clearly shows a commitment by the accountable executive beyond lip service, but actually promising to commit the resources necessary to meet the safety objectives, not just put it on a sign or give a speech to all the new hires. A possible statement might be, "As the accountable executive, I am committed to provide the resources necessary to establish and maintain a safety management system to fulfilled the organizations safety objectives which are:…"

- **A safety reporting policy that defines requirements for employee reporting of safety hazards or issues.** This requirement turns the commitment upside down, from the bottom up. It indicates to all the stakeholders that they are needed to establish and maintain an SMS. The safety reporting policy itself does not need to be in the safety policy statement, but it must at least show the commitment by the accountable executive as well as give clear direction as to where the complete safety reporting policy is located.

- **A policy that defines unacceptable behavior and conditions for disciplinary action.** This stems for the need of a Just culture; a culture where everyone knows what the expectations are and that contains a commitment that employees will be treated fairly. Normally a commitment by the accountable executive is made within the safety policy statement and a more detailed description is located in the SMS manual under the safety policy section.

- **An emergency response plan (ERP) that provides for the safe transition from normal to emergency operations in accordance with the requirements of 14 CFR §5.27.** As with the other components of the safety policy statement, details do not need to be within the statement, but there must be additional guidance referenced that meets all the criteria. Under a safety program, many companies will have most of these requirements complete. Safety programs are generally reactive in nature, including an emergency plan to react to an accident or incident. Usually is it some sort of checklist identifying the entity to be called in case of an emergency. If it is a more mature safety program it may have a flow chart identifying who is responsible for what actions in case of an emergency. SMS formalized this, requiring key components under §5.27 as a minimum acceptable that has been approved by the accountable executive. These minimums are:
  ▷ Delegation of emergency authority throughout the certificate holder's organization;
  ▷ Assignment of employee responsibilities during the emergency; and
  ▷ Coordination of the certificate holder's emergency response plans with the emergency response plans of other organizations it must interface with during the provision of its services.

It is important to verify that when key decision makers are unavailable to fulfill their responsibilities, the certificate holder has position proxies or a backup plan to maintain the affected processes. The FAA will want to ensure that the organization's back-up strategy (people and processes) will work.

A certificate holder's ERP documentation should identify substitutes for those that must participate in emergency activities and are then unavailable to perform normal duties. Here are a few questions that an organization should ask, as well as what the FAA certainly will ask:

1. How the person is notified of their additional duties;
2. That, as a proxy, they have the competencies (training) to perform the additional duties; and
3. That the person is knowledgeable of these duties or can identify appropriate guidance required for performance.

The organization will need to ensure they are coordinating its ERP with other organizations' emergency plans. How will the local law enforcement be notified? Who will do it? Who will they contact? Is there evidence through meeting minutes, documents or other means, such as contracts, that validate communication with various organizations?

An organization will want to conduct a "tabletop" exercise to demonstrate the effectiveness of the documented processes, and should include other organizations that may be affected. The tabletop exercise should further identify individual's duties and responsibilities in any unlikely event. Many airport conduct yearly practice drills or tabletop exercises. If possible, contribute or lead such an event in order to practice your ERP and determine its true effectiveness before the FAA asks. **The accountable executive described in 14 CFR §5.25 must sign the safety policy statement.**

THE SAFETY POLICY

All items in the safety policy statement must be a part of the overall organizational safety policy, but your safety policy includes many required items that not specifically in the safety policy statement signed by the accountable executive.

The organization's safety policy items are normally kept in a central SMS manual or in other standard operation procedures. The current safety policy documentation should be clearly documented, employees should know where to find it, and should be trained to use it. Best practice would be to have one safety policy section in an SMS manual with the safety policy statement giving a summary of all the required items. Within it should be the remaining safety policy, providing additional details regarding the safety policy statement as well as addressing the other requirements outlined in Part 5 and/or the SMSVP guide.

Additional items required within a SMS safety policy are as follows:

1. Ensuring any reference in the safety policy statement is substantiated and consistent throughout the remaining safety policy. If your policy statement indicated a different manual where policy, procedures, or processes are found, it must be clearly documented.
2. Ensuring the certificate holder's process regarding effectively communicating its safety policy at all levels of the organization is communicated to existing, new, and temporary employees, as applicable.

3. All levels of management must be aware of their responsibility and accountability for safety in their organization. Individual managers are responsible for developing, implementing, and maintaining SMS processes within their technical areas. Members of management must be aware of their accountability and competence at:
   - Hazard identification and safety risk assessment;
   - Assuring the effectiveness of safety risk controls;
   - Promoting safety; and
   - Advising the accountable executive on the performance of the SMS and any need for improvement.

This is a significant variance from a safety program where you have a safety director or vice president of safety who is responsible for protection, while the rest of the organization is concerned with production. The most effect method of change to encourage all levels of management and all personnel to be concerned with safety is to make it a part of their performance evaluations.

Different organizations have different evaluation processes, but the important aspect is that personnel must be evaluated not only on their production, but also on their protection or safety efforts.

It is human nature to think only about production. For example, a manager might think, "At the end of the year, I will be told I am doing a good job or a poor job based on production, therefore, I will focus on production throughout the year." Most organizations are not familiar with measuring an individual's safety efforts, all the metrics used to measure performance are things that relate to increasing financial status, making production more effective, or the maintaining relationships with those we work with. Rarely do we have safety metrics to measure safety performance. The key to an effective safety metric is it must be something, you as the employee being evaluated, can control. If I am the chief pilot and my safety metric is the number of accidents, it will not truly measure my safety efforts because my ability to control all the aspects of an accident are unattainable. Conversely, if I am evaluated on my safety efforts in regard to expecting safety risk assessments to be completed and my prompt follow up with risk mitigation strategies that were approved, as a chief pilot I can control that and prioritize that. In the scenario for discussion exercise of this chapter (see page 127) you have an opportunity to consider various management positions and identify what safety metrics would be appropriate.

Accountable Executive

While most individuals in your organization are do not have specific responsibilities identified in Part 5, the accountable executive and a manager designated by the accountable executive are addressed. As an accountable executive the following items must be accomplished and documented in the safety policy.

1. **The accountable executive must ensure that the SMS is properly implemented and performing in all areas of the certificate holder's organization.** This is done by an active involvement by the accountable executive in safety decisions. How this is done is at the discretion of the organization, but when and how it is done must be documented.

2. **The accountable executive must develop and sign the safety policy of the certificate holder.** This is to require direct involvement for those key management personnel including the accountable executive who are ultimately responsible for safety.

3. **The accountable executive must communicate the safety policy throughout the certificate holder's organization.** The accountable executive sets the culture in regard to commitment the organization truly has to managing safety. A written safety policy is part of the communication, but the efforts and decisions made to actually support the safety policy through resources and personnel will communicate much more than the written word.

4. **The accountable executive must regularly review the safety policy to ensure it remains relevant and appropriate to the certificate holder.** These are key words as you move forward with SMS are, relevant and appropriate. If your safety policy, and SMS in general, is to be effective you must review your policy, your decision-makers must be a part of that review process and you must document that when and how it is to be reviewed.

Designation of Management Personnel

The accountable executive must designate sufficient management personnel who, on behalf of the accountable executive are responsible for the following:

1. Coordinating implementation, maintenance, and integration of the SMS throughout the certificate holder's organization.
2. Facilitating hazard identification and safety risk analysis.
3. Monitoring the effectiveness of safety risk controls.
4. Ensuring safety promotion throughout the certificate holder's organization as required under Part 5 Safety Promotion Subpart E.
5. Regularly reporting to the accountable executive on the performance of the SMS and on any need for improvement.

The responsibilities of management personnel must also be documented. These key individuals will most likely be your vice presidents and directors, who will be given further responsibility regarding acceptable risks. The term used by the FAA for these types of individuals is **process owners**. Process owners are these key safety management personnel who oversee specific areas of operations, such as, director of flight operations or director of maintenance. Process owners may further delegate duties to **process managers** such as a chief pilot or maintenance training manager. Process managers are individuals assigned by a process owner who can accept the appropriate level of risk within their area of responsibility. Process managers' duties may include:

- Hazard identification;
- Safety risk assessment, and risk, acceptance;
- Evaluating the effectiveness of safety risk controls;
- Promoting safety; and
- Submitting performance reports to the key safety management/process owners on SMS performance.

Note: "Accountability" as used here, refers to active management and line employee involvement and action in managing and maintaining safety performance. A certificate holder defines accountability by ensuring that each of its management and line employees is aware of his or her specific role within the SMS and actively participates in carrying out his or her SMS related duties. Once the accountabilities for these employees have been defined, 14 CFR Part 5, Safety Promotion, Subpart D, requires that these accountabilities be communicated throughout the organization.

SAFETY OBJECTIVES

AC 120-92 states, that safety objectives will fall into a number of categories, including:

1. Compliance with regulations. Compliance with all regulations is an expectation for all certificate holders and assurance of such compliance is an explicit requirement of the SMS (refer to §5.3(b)).
2. Milestones for implementation of safety-related programs or initiatives.
3. Reduction of error or incident rates.
4. Tracking of safety events. Certain events such as ground damage, pilot deviations, weight and balance errors, or maintenance errors may be targets for safety objectives and associated tracking and action. One caution with these types of measures is not to lose focus on risk factors that may be associated with potentially more serious events.

These safety objectives are critical to your SMS success. As your organization develops them, ensure they are specific enough that they can be measured and they are objectives that can realistically be attained. Developing these objective take directly involvement from the accountable executive as well as the designated safety management personnel within your organization. This would include the director of safety, as well as any other individual who is has been designed by the accountable executive, such as the director of maintenance.

These objectives may be both short and long term in nature. Milestones for implementation or tracking critical safety events may be short while other items may be more long term consistent with your business strategies. **Remember, SMS should align with your business goals and strategies, not counteract them as the traditional protection and production scale seems to suggest.**

Figure 9-1 on page 124 is an example of beginning your development of safety objectives for a Part 141 pilot school.

SAFETY PERFORMANCE

Safety Performance: Realized or actual safety accomplishments that are relative to the organization's safety objective(s).

The accountable executive is defined as a key leadership individual in the organization's business tier that has ultimate authority over safety operations and organizational resources. As a

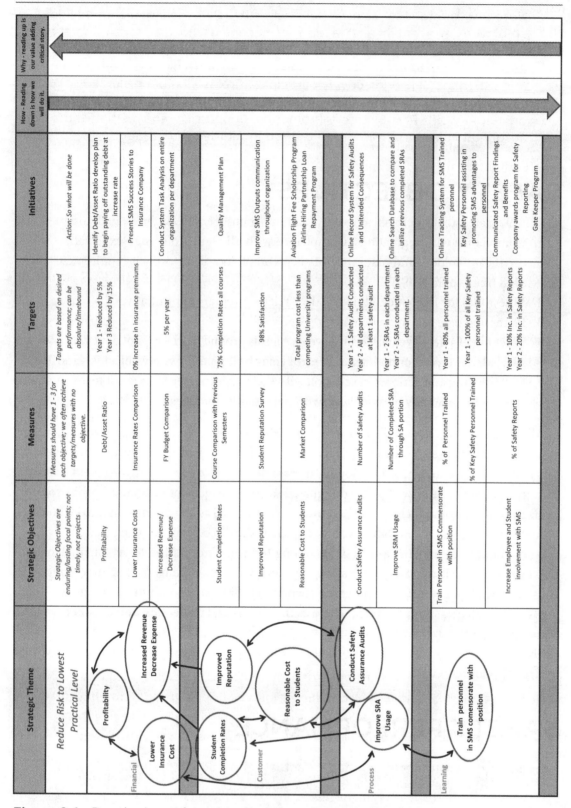

Strategic Theme	Strategic Objectives	Measures	Targets	Initiatives	How - Reading down is how we will do it.	Why - reading up is our value adding critical story.
Reduce Risk to Lowest Practical Level	*Strategic Objectives are enduring/lasting focal points; not timely, not projects*	*Measures should have 1 - 3 for each objective; we often achieve targets/measures with no objective.*	*Targets are based on desired performance; can be absolute/timebound*	*Action: So what will be done*		
	Profitability	Debt/Asset Ratio	Year 1 - Reduced by 5% Year 3 Reduced by 15%	Identify Debt/Asset Ratio develop plan to begin paying off outstanding debt at increase rate		
	Lower Insurance Costs	Insurance Rates Comparison	0% increase in insurance premiums	Present SMS Success Stories to Insurance Company		
	Increased Revenue/ Decrease Expense	FY Budget Comparison	5% per year	Conduct System Task Analysis on entire organization per department		
	Student Completion Rates	Course Comparison with Previous Semesters	75% Completion Rates all courses	Quality Management Plan		
	Improved Reputation	Student Reputation Survey	98% Satisfaction	Improve SMS Outputs communication throughout organization		
	Reasonable Cost to Students	Market Comparison	Total program cost less than competing University programs	Aviation Flight Fee Scholorship Program Airline Hiring Partnership Loan Repayment Program		
	Conduct Safety Assurance Audits	Number of Safety Audits	Year 1 - 1 Safety Audit Conducted Year 2 - All departments conducted at least 1 safety audit	Online Record System for Safety Audits and Unitended Consequences		
	Improve SRM Usage	Number of Completed SRA through SA portion	Year 1 - 2 SRAs in each department Year 2 - 5 SRAs conducted in each department.	Online Search Database to compare and utilize previous completed SRAs		
	Train Personnel in SMS Commensorate with position	% of Personnel Trained	Year 1 - 80% all personnel trained	Online Tracking System for SMS Trained personnel		
		% of Key Safety Personnel Trained	Year 1 - 100% of all Key Safety personnel trained	Key Safety Personnel assisting in promoting SMS advantages to personnel		
	Increase Employee and Student involvement with SMS	% of Safety Reports	Year 1 - 10% Inc. in Safety Reports Year 2 - 20% Inc. in Safety Reports	Communicated Safety Report Findings and Benefits Company awards program for Safety Reporting Gate Keeper Program		

Figure 9-1 Developing safety objectives for a Part 141 pilot school.

result, it is important that the accountable executive is aware of safety performance data and information collected from the system so that he/she may direct any necessary actions and/or resources to support safety initiatives.

Therefore, SMS process requires defined accountability for achieving safety performance objectives within the organization's safety policy. To do that the certificate holder must conduct assessments of its safety performance against its safety objectives, which include reviews by the accountable executive, to:

1. Ensure compliance with the safety risk controls established by the certificate holder;
2. Evaluate the performance of the SMS;
3. Evaluate the effectiveness of the safety risk controls and identify any ineffective controls;
4. Identify changes in the operational environment that may introduce new hazards; and
5. Identify new hazards.

Lastly, upon completion of this assessment, if ineffective controls or new hazards are identified during this safety performance review, the certificate holder must use the SMS safety risk management process to further mitigate risk.

Safety performance reviews are led by the accountable executive and should include the process owners at a minimum. For example, each department may have a process owner. During a safety performance review meeting, each process manager identifies hazards and risk within their department, what they are doing to reduce the risk, and discusses safety risk assessments. The accountable executive leads the conversation and at times, makes risk decisions. Most importantly, the safety performance is measured directly with the safety objectives. During this meeting, the accountable executive reviews the safety policy and gives each process manager an opportunity to discuss any safety concerns for follow up on any safety actions assigned by the accountable executive.

Operationally, organizations have been having this type of meeting for decades, but as a safety performance review the concept is novel. Depending on your operation, this type of meeting may be assimilated into any meeting where the same individuals already meet. Organizations should establish, at a minimum, quarterly meeting to specifically address organizational safety performance and safety objectives. With this meeting, it is critical that meeting notes are taken, action items are recorded, and a standard template is used to ensure the safety objectives are central to the discussion and the decisions enacted. It is often difficult to identify directed actions resulting from meeting minutes unless a template is used to list defined actions to be carried forward to the next safety management review. Using this technique removes the guesswork associated with deciphering discussions contained in meeting minutes. The SMSVP strongly encourages certificate holders to use a template or actions table for meeting minutes.

At your final audit with the SMSPO you will be evaluated while conducting a safety performance review with the accountable executive. At a minimum the FAA will want three key questions answered:

1. Does the certificate holder have documentation showing that the accountable executive is periodically reviewing and assessing the organization's safety management performance?
2. Does the certificate holder have documentation showing that the accountable executive directs actions to address substandard safety performance?

3. Does the certificate holder have documentation showing the directives of the account-able executive are tracked and reported upon at the next regular review or as required?

SCENARIO FOR DISCUSSION

SMS Policy Statement Exercise

Listed below is the safety policy statement written by Snyder Safe Airline. As a safety officer within your organization, you have been asked to review and ensure it meets the requirements of a safety policy statement under Part 5. Analyze if it meets the SMS requirements and if not, write in the space provided that which must be added or changed to clearly meet SMS requirements.

Snyder Safe Airline SMS Policy Statement

To prevent aviation accidents and incidents our organization will maintain an active safety management system.

I support the open sharing of information on all safety issues and encourage all employees to report significant errors, safety hazards or concerns.

I pledge that no staff member will be asked to compromise our safety standards to "get the job done."

Safety is a corporate value of this company, and we believe in providing our employees and customers with a safe environment.

All employees must comply with this policy.

Our overall safety objective is the proactive management of identifiable hazards and their associated risks with the intent to eliminate their potential for affecting aviation safety, and for injury to people and damage to equipment or the environment.

To that end, we will continuously examine our operation for these hazards and find ways to minimize them.

We will encourage hazards and incident reporting, train staff on safety management, document our findings and mitigation actions and strive for continuous improvement.

Ultimate responsibility for aviation safety in the company rests with me as the President. Responsibility for making our operations safer for everyone lies with each one of us—from managers to front-line employees.

Each manager is responsible for implementing the safety management system in his or her area of responsibility, and will be held accountable to ensure that all reasonable steps are taken.

Dr. Robert Snyder
SSA President

REVIEW QUESTIONS

1. Describe the difference between the safety policy statement and the safety policy.

2. Define safety objectives and explain their importance to SMS.

3. Describe how the job responsibilities of a director of flight operations might change under an SMS.

4. Define accountable executive.

5. List three responsibilities of an accountable executive.

6. What is safety performance?

7. Compare and contrast the positions of process managers and process owners.

8. What items must be included in a safety policy statement?

9. Explain the significance of safety objectives.

10. Explain why it is critical to have participation from the accountable executive and key management personnel when developing safety objectives.

11. Create a safety policy statement that meets all the requirements of 14 CFR Part 5. Use your own company as the model or identify an actual company that closely aligns with your field of study. If you were educated as an Airport manager identify an airport in which you can develop a safety policy statement that would meet SMS Part 5 requirements. Remember this is a statement written by the accountable executive and will most likely be prominently displayed in your future SMS manual or Safety website. As required by the 14 CFR Part 5 ensure your safety policy statement also includes your key safety objectives for your company.

CHAPTER 10
Safety Risk Management

OBJECTIVES

- Understand how the SRM component of SMS provides a decision-making process for identifying hazards and mitigating risk.
- Explain how the SRM concepts are based on a thorough understanding of your organization's systems and operating environment.
- Describe how the SRM component is the organization's way of fulfilling its commitment to consider risk in their operations and to reduce it to an acceptable level.
- Apply the SRM concepts to complete a safety risk assessment.

KEY TERMS

- Five "M" Model
- Risk Control

INTRODUCTION

SRM is a formal system of hazard identification and management. SRM is fundamental in controlling an acceptable level of risk. A well designed risk management system describes operational processes across department and organizational boundaries, identifies key hazards and measures them, methodically assesses risk, and implements controls to mitigate risk.

Understanding the hazards and inherent risks associated with everyday activities allows the organization to minimize unsafe acts and respond proactively, by improving the processes, conditions and other systemic issues that lead to unsafe acts. These systemic/organizational elements include training, budgeting, procedures, planning, marketing and other organizational factors known to play a role in many systems-based accidents. In this way, safety management becomes a core function and is not just an adjunct management task. It is a vital step in the transition from a reactive culture, one in which the organization reacts to an event, or to a proactive culture, in which the organization actively seeks to address systemic safety issues before they result in an active failure. The fundamental purpose of a risk management system is the early identification of potential problems. The risk management system enhances the manner in which management safety decisions are made.

SRM: PROACTIVE AND REACTIVE HAZARD AND RISK MITIGATION

Although aviation organizations use SRM as a formal safety and risk assessment process, its philosophy is easily understood outside of the technical realm of aviation. For example, a person performs SRM each time he or she crosses the street. The individual identifies hazards (cars passing), analyzes and assesses the risk (potential to be struck and severity if he or she is), and explores mitigations (looking both ways for traffic and/or heeding pedestrian signals) to reduce the perceived risk to an acceptable level before proceeding.

It is necessary to make the approach to managing safety risk into a formalized, objective process. This helps ensure the effective management and mitigation of safety hazards and risk. SRM provides a means to:

- Identify potential hazards and analyze and assess safety risk in ATO operations and NAS equipment;
- Define mitigations to reduce risk to an acceptable level;
- Identify safety performance targets to use as a benchmark for the expected performance of mitigations; and
- Create a plan that an organization can use to determine if expected risk levels are met and maintained.

These concepts can be put into a logical flowchart as shown in Figure 2-7 on page 38.

System Description and Task Analysis

The first step in SRM is system description and task analysis. Here, the analysis need only to be as extensive as needed to understand the processes in enough detail to develop procedures, design appropriate training curricula, to identify hazards, and to measure performance. Establish the context. This is the most significant step of the risk process. In SRM, the first step, is used to understand the aspects of the operation that might cause harm. In most cases, hazard identification flows from this system analysis. Hazard identification requires you to ask: what hazards exist in the operational environment? What are the human factors issues of the operation (e.g., workload, distraction, fatigue, or system complexity)? What are the limitations of the hardware, software, procedures, etc.? It defines the scope and definition of the task or activity to be undertaken, the acceptable level of risk is defined, and the level of risk management planning needed is determined.

Hazard Identification

Next, we look at the processes and play "What if?" What could go wrong with our processes, under typical or abnormal operational conditions that could be considered hazardous? Here we identify the hazards and risks. While the diagram above depicts that processes as distinctly defined components, in practice they flow from one to the other. For example, in a careful discussion of how a system currently works (system description (analysis)), hazards will often become evident. Thus, the hazard identification step has also been at least partially accomplished. Identification of what could go wrong and how it can happen is examined, hazards are also identified and reviewed, and the source of risk or the potential causal factors are also identified.

Risk Analysis

Based on the analysis in the hazard identification step, we determine the injury and damage potential of the events related to the hazards in terms of likelihood of occurrence of the events and severity of resulting consequences. The process then progresses into an analysis of the potential consequences of operation in the presence of the identified hazards. Determine the likelihood and consequence of risk in order to calculate and quantify the level of risk. A good tool for this process is the reporting system for information gathering technique. Determining the frequency and consequence of past occurrences can help to establish a baseline for your risk matrix. Each organization will have to determine their definition of severity according to its individual risk aversion.

Risk Assessment

Risk assessment is a decision step based on combined severity and likelihood. Is the risk acceptable? Where potential severity is low or if likelihood is low or well mitigated with existing controls, we may be done and ready for operation. If not, we'll need to move to the next step, designing risk controls. Here we determine whether the risk is acceptable or whether the risk requires prioritization and treatment. Risks are ranked as part of the risk analysis and evaluation step. This culminates in an assessment of the acceptability of operating with these hazards (risk assessment), or whether or not the risk of such operations can be mitigated to an acceptable level (**risk control**). For this reason, operational managers must be the ones who are accountable for these decisions.

Risk Control

If the risk assessment is unacceptable, we will need to design risk controls. Most often, these entail either new processes or equipment, or changes to existing ones. We then look at the system with the proposed control in place to see if the level of risk is now acceptable. We'll stay in this design loop until we determine that the proposed operation, change, etc. can be mitigated to allow operations within an acceptable level of risk. Adopt appropriate risk strategies in order to reduce the likelihood or consequence of the identified risk. These could range from establishing new policies and procedures, reworking a task, or a change in training, to giving up a particular mission or job profile.

Monitor and Review

This is required during all stages of the risk process. Constant monitoring is necessary to determine if the context has changed and the treatments remain effective. It should not be surprising to find at this time that there are still things that might not have been considered or that there are changes over time in the operational environment, requiring a return to SRM. Thus, the SRM and SA processes operate in a continuous exchange. In the event the context changes, a reassessment is required.

Risk Matrix

The risk assessment matrix is a useful tool to identify the level of risk and the levels of management approval required for any risk management plan. There are various forms of this matrix, but they all have a common objective to define the potential consequences and/or severity of the hazard versus the probability or likelihood of the hazard.

To use the risk assessment matrix effectively it is important that everyone has the same understanding of the terminology used for probability and severity. For this reason, definitions for each level of these components should be provided. Figure 2-4 on page 23 shows a sample risk matrix that could be used by an aviation organization.

SYSTEMATIC METHOD OF CONDUCTING AN SRA

Your organization will need to develop a risk assessment process to understand the critical characteristics of your systems and operational environment, and apply this knowledge to identify hazards, analyze and assess risk, and design risk controls. The SRM process will be applied to:

- Initial design of systems, organizations, and/or products;
- The development of operational procedures;

- Hazards that are identified in your SMS manual; and
- Planned changes to operational processes.

List all members participating in the risk assessment. This is important to document the people who conducted the risk assessment. This is critically important in understanding "why" when the risk assessment undergoes its annual and recurring reviews.

STEP ONE: IDENTIFY HAZARDS

The first step is an identify and generally describe the hazard. Enter one hazard per sheet. The objective is to identify those things most likely to have a negative impact on the mission. Remember to identify the ultimate cause of something that might go wrong. As an example, cutting your finger is a risk, but what was the cause of cutting your finger? The knife. The knife is the hazard, cutting yourself is the risk. Sliding off the departure end of the runway is a risk, but what was the cause of sliding off the end of the runway? It might be hydroplaning. What is the cause of hydroplaning? Standing water on the runway. What is the cause of standing water on the runway? Poor runway grooving. The ultimate hazard is poor runway grooving.

The Five "M" Model can be a good source for identifying hazards. Hazards often come from man, media (environment), the machine, management, and the mission. In risk assessment, the operator is always an essential element. Ultimately, we need to anticipate the risks associated with the human who operates the machine within a media under management criteria.

Man

A safety risk assessment needs to account for the human component responsible for the safe operation of the aircraft or system. Successful safety risk assessments requires the team to probe beyond the possibility of human failure so as to be able to determine the underlying factors that contribute or lead to this human failure (negative consequence). The question "why" arises a lot during investigation of the operator at the time of a negative consequence.

For example, what if:
- An individual is not properly trained in how to cope with the situation that could lead to an accident?
- Nobody is responsible for dealing with the training deficiency?
- The individual is not mentally or physically capable of responding properly?
- A human failure occurs due to a self-induced state such as alcohol intoxication or fatigue?
- The individual was not given adequate operational information on which to base decisions?
- Nobody is responsible for providing the proper information?
- The individual can become distracted to the point that he/she is not paying proper attention to their duties?

The answers to each of such questions are vital for the effective measures to be put in place so as to prevent the accident from occurring. Some of these human elements are:

Selection

Is the right person, emotionally/physically, trained? Select the right person for the right job. Behaviors are usually considered normal if limited to one, two, or three (max) tendencies. We all probably exhibit one or two of these tendencies.

- Accident prone
- Antagonized by authority
- Anxious, tense, panicky
- Blame-avoidant, always has an excuse
- Cannot concentrate
- Careless or frivolous
- Channelized attention
- Chronically indecisive
- Circadian rhythm—working between midnight and 6 am
- Constantly in need of guidance
- Does not recognize limitations of co-workers
- Does not recognize own limitations
- Does not think well on their feet
- Easily influenced or intimidated
- Event proficiency
- Fatigue, mental—difficult job given at the end of the shift
- Fatigue, physical
- Feels inadequate
- Follows procedures, not techniques
- Frustrated and discontented
- Helpless
- Highly competitive
- Holds grudges, harbors resentment
- Hypersensitive
- Immature, foolhardy, impetuous
- Improper perception
- Intolerant, impatient
- Irresponsible
- Irritable and cantankerous
- Lack of confidence
- Lacks common sense
- Lacks self-discipline
- Must continually prove themselves
- Not aware of the surroundings
- Not a "motivated" person—does the minimum acceptable
- Not disciplined, known for cutting corners when under pressure to perform
- Over confident
- Overconfident
- Overly sensitive to criticism
- Peer pressure
- Poor communication skills
- Predisposed to alcohol or drugs
- Prone to habit patterns
- Resists authority
- Selfish or self-centered
- Sleep deprived—many days without enough sleep
- Stress
- Task saturated—too much to do, with too little time
- Unrealistic goals
- Workload/pressure—too much to do, with too little time

Note: It is very important for you to understand why people take risks. It is our objective to understand these principles and plan accordingly.

Why do people take risks?

- Undisciplined
- Time pressure
- Highly motivated
- Supervisory pressure
- Ignorant or untrained
- Action is tolerated
- Correct equipment not available

Results vs. Method

Because we have a tendency to reward results rather than method it can lead to the situation where risky behavior is tolerated, even encouraged, as long as it gets the job done. If an accident occurs, though, watch out.

Hazardous Attitudes

- Anti-authority—don't tell me what to do. Don't tell me how to do my job.
- Impulsivity—we have to get it done right now. The risk gets higher the longer we wait.
- Invulnerability—I know my stuff, I've done this before. I will not cause an accident.
- Machismo—I am better then you guys. I'll show you how a true master gets it done.
- Resignation—oh well, this is a lose-lose situation. I am just a pawn. They will never listen to me.

Personal Behavior Patterns of People Exhibiting Good Risk-Taking Characteristics

- Well-balanced, well-controlled
- Ingrained sense of responsibility
- Aware of own weaknesses and limitations
- Moderation
- Healthy and realistic outlook and goals
- Positive attitudes
- Satisfactory interpersonal relations
- Mature
- Kindly and tolerant attitudes toward others
- Healthy and realistic outlook and goals
- Well-developed social and civic conscience
- Reasonably intelligent
- Adequate control over impulses and emotions
- Ingrained sense of responsibility
- Accurately able to assess situations and act
- Contented, adaptable, accepting

Media (Environment)

This is the environment with which we work. This includes climate, terrain, and noise/distraction and runway environment. These external, largely environmental, forces vary and must be considered when assessing risk:

- Natural environment. Climactic: Temperature, seasons, precipitation, aridity, wind, visibility.
- Artificial environment. Operational, routes, surfaces, terrain, vegetation, obstructions, and constrictions.
- Hygienic. Vent, noise, toxicity, corrosives, dust, and contaminants.
- Vehicular/Pedestrian. Paved, gravel, dirt, ice, mud, dust, snow, sand, hilly, curvy.

Machine

The machine category encompasses the aircraft, tug, fuel truck, maintenance vehicle, power cart, and more. The machine category includes the equipment's design, its maintenance history and performance, its maintenance technical orders and its user perception.

Technology has enabled great advances to the aviation industry. Through automation, human mental workload has dropped significantly and human productivity increased. When machine and computer become more complicated and replace more human jobs, problems with human limitations handling the technology can surface. Modern aircraft designs are revised to further reduce the effect of any of these hazards. Good design not only seeks to make system failure unlikely, but it also ensures that, if a system does fail, one single failure will not result in an accident.

On a multi-million dollar airplane this hazard reduction could include:

- Design: engineering and user friendly (ergonomics).
- Maintenance: Training, time, tools, parts.
- Logistics: supply, upkeep, and repair.
- Technical data: clear, adequate, useable, and available.

Management

Management provides the enforcement and establishment of standards, procedures and controls. It drives the interaction between man, media, machine, and mission. There is significant overlap between man, machine, media and mission because these elements interrelate directly, but the critical element is management. Any breakdown within the man, machine, mission or media must be viewed as an effect of management performance. When outcome fails to meet anticipated goals, these Five M's must be thoroughly reassessed. Management is the controlling factor in defining the process of either production success or failure.

Management dictates the process by defining:

- Standards,
- Procedures, and
- Controls.

Inadequate Supervision

- Failed to provide guidance.
  - ▷ Boundaries for undisciplined employee.
  - ▷ Boundaries for highly motivated.
  - ▷ Everyone does the job differently.
- Failed to provide operational doctrine.
- Failed to provide oversight.
- Failed to provide training.
- Failed to track qualifications.
- Failed to track performance.
- Failed to provide clearly defined objectives.
- Failed to provide incentives for meeting goals/objectives.

Failed to correct a known problem

- Failed to correct a document in error.
- Failed to identify an at-risk employee.
- Failed to initiate corrective action.
- Failed to report unsafe conditions.
- Failed to discipline intentional violations.
- Failed to encourage the reporting of any unsafe condition/hazard.
- Failed to demand compliance.

Planned Inappropriate Operations

- Failed to provide correct data.
- Failed to provide proper equipment.
- Failed to provide proper manning at the right time.
- Failed to provide adequate time for the task.
- Failed to provide adequate opportunity for rest.
- Task not in accordance with rules/regulations.
- Production-based appraisals/measurements.
- Authorized an unsustainable high operational tempo.
- Established high production quota.

Supervisory Violations

Authorized unnecessary risk/hazard.

- Failed to enforce all the rules and regulations.
- Authorized unqualified person for the job.
- Forced employee to choose between production or safety.
- Pressured employees to "be loyal and accept an unnecessary risk."

Resource Management

- Failed to provide proper monetary resources
  - ▷ Excessive cost cutting
  - ▷ Lack of funding
- Equipment/facility resources
  - ▷ Poor design
  - ▷ Purchasing of unsuitable equipment

Culture

- Failed to encourage a learning culture
- Failed to encourage an informed culture
- Failed to encourage a flexible culture
- Failed to encourage a just culture
- Failed to encourage a reporting culture

Mission

As an example, a flight school has a mission to train pilots under 14 CFR Part 141. The flight school is aware of the basic risk of conducting cross-country training flights during bird migratory season. With their Part 141 Pilot Schools mission, they understand the need to conduct night cross-country flights, even during bird migratory season. This mission to train future pilots results in the conscious decision to accept a few pre-defined risks, with proper mitigation. In our case we can apply risk management techniques during bird migratory season: day flights are better than night flights. Climbing to avoid birds is better than descending. High altitudes are better than low altitudes. Slower airspeeds are better than fast airspeeds. Flights over the desert are better than flights over terrain with lots of open water.

STEP 2: HAZARD CONSEQUENCE DESCRIPTION

The objective is to document the end result assuming the hazard event occurs (do not take into account the probability, at this point). Typical responses include some type of injury, damage to equipment, death, loss of aircraft, or loss of certificate (organization and/or the individual). Referring to the example above, the hazard is poor runway grooving. The consequences of poor runway grooving could be: standing water on the runway, causing hydroplaning, resulting in a loss of directional control, sliding off the end of the runway going into the ocean. Loss of directional control during takeoff could mean sliding off the side of the runway at a high rate of speed, and striking runway lights at 70 knots. Loss of directional control on landing could mean departing the side of the runway, going into a steep drainage ditch at 60 knots.

STEP 3: IDENTIFY YOUR HAZARD MITIGATION CONTROLS

Describe the existing control measures currently used to mitigate hazards that are already being used by your organization. Summarize how these existing controls are implemented. As needed, specify the: who, what, where, when, and how for each control. As an example, your safety policies and procedures manual states that fixed-wing operations cease whenever the runway crosswind exceeds 30 knots, and solo fixed-wing flights cease when the runway crosswind exceeds 15 knots.

STEP 4: PRE-MITIGATION RISK ASSESSMENT

Use the criteria outlined in the risk assessment matrix (see Figure 2-4 page 23) to determine the severity, probability, and risk taking into account all of the exiting controls used by the organization.

STEP 5: ANALYZE THE RISK CONTROL MEASURES YOU PLAN TO USE

Investigate specific strategies and tools that reduce, mitigate, or eliminate the risk. Effective control measures reduce or eliminate one of the three components (probability, severity, or exposure) of risk. The following options assist in identifying potential controls:

Reducing Risk

The overall goal of SRM is to plan missions or design systems that do not contain hazards. A proven order of precedence for dealing with hazards and reducing the resulting risks is:

- Design (engineer) for minimum risk;
- Incorporate safety devices;
- Provide warning devices;
- Install guards; and
- Improve procedures (task design) and training.

Design Selection/Engineering

- Plan or design for minimum risk. Design the system to eliminate hazards. Without a hazard there is no probability, severity or exposure. As an example, design connectors that do not allow components to be wired backwards. The hazard is now gone
- Eliminate hazards through design selection. If we need to fly in icing conditions, take the aircraft that is approved for flight in known icing. Ideally, the hazard should be eliminated by selecting a design or material alternative that removes the hazard altogether.
- Reduce risk through design alteration. If adopting an alternative design change or material to eliminate the hazard is not feasible, consider design changes that reduce the severity and/or the probability of the mishap potential caused by the hazard(s). As an example, engine fire suppression systems reduce the probability and severity of an uncontrolled engine fire.
- Incorporate engineered features or devices. If mitigation of the risk through design alteration is not feasible, reduce the severity or the probability of the mishap potential caused by the hazard(s) using engineered features or devices. In general, engineered features actively interrupt the mishap sequence and devices reduce the risk of a mishap. As an example, a tripped circuit breaker actively interrupts the risk of an inflight electrical fire.

Incorporate Safety Devices

- Where design alternatives, design changes, and engineered features are not feasible, incorporate safety devices to mitigate the severity of the mishap potential. Typical strategies reduce the risk via the use of fixed, automatic, or other safety features or devices, and make provisions for periodic functional checks of safety devices.
- These safety features or devices are currently available and do not need to be designed or engineered. They usually do not affect the mishap probability; they reduce the severity.
  - ▷ For example, seat belts and air bags do not lower the probability of an accident, however they reduce the severity of injuries.

> ▷ An automatic "low altitude" detector in a surveillance system.
> ▷ A refueling nozzle grounding circuit to eliminate static discharge during refueling operations.

Provide Warning Devices

- If engineered features and devices are not feasible, and the use of safety devices do not adequately lower the severity or probability of the mishap potential caused by the hazard, include detection and warning systems to alert personnel to the presence of a hazardous condition or occurrence of a hazardous event.
- The warning must be provided in time to avert the hazard effects.
- The applicable warnings and their application are designed to minimize the likelihood of inappropriate human reaction and response.
- Insert a warning in the operator's manual.
- "Engine Failure" light in a helicopter.
- Flashing warning on a radar screen.
- Incorporate signage, procedures, training, and PPE.
  - ▷ Signage includes placards, labels, signs and other visual graphics.
  - ▷ Procedures and training should include appropriate warnings and cautions. Procedures may prescribe the use of PPE.
  - ▷ For hazards assigned catastrophic or critical mishap severity categories, the use of signage, procedures, training, and PPE as the only risk reduction method should be avoided.
- Warning devices may be used to detect an undesirable condition and alert personnel.
  - ▷ Examples include: traffic collision advisory system (TCAS), enhanced ground proximity warning system (EGPWS), wind sheer alerting system, autopilot disengaged aural, landing gear warning horn, engine fire warning bell.

Guard

- Where it is impractical to eliminate hazards through engineering or design selection, or adequately reduce the associated risk with safety and warning devices, guards should be used.
- Create and install physical guards to identify and sanitize the area from the hazard. This guard can be a physical barrier designed to keep personnel out of the danger area.
  - ▷ Another example would be to install a safety barrier between the fan blades of a jet engine and the bottom of the wing. This barrier would stop failed fan blades from putting holes in the bottom of the wing.
  - ▷ One additional example is to install a guard between critical hydraulic lines and the fan blades of a jet engine. If the fan blades fail, the guard will protect the hydraulic lines.
- You can also put a safety guard on the source. As an example, you could engineer an engine cowling that would contain any failed fan blades inside the jet engine.
- You can put protection on the human. This is often called personal protection equipment (PPE). Steel-toed boots, Kevlar bullet proof vests, and safety glasses are common examples of PPE.

- You can also raise the threshold, or "harden" any of the above when identified as inadequate protection.

Improve Procedures (Task Design) and Training

- Where it is impractical to eliminate hazards through design selection or adequately reduce the associated risk with guards, or safety and warning devices, procedures and training should be used.
  - ▷ The use of guards and/or a warning system by itself may not be effective without training or procedures required to respond to the hazardous condition.
  - ▷ The greater the human contribution to the functioning of the system or involvement in the mission process, the greater the chance for variability.
  - ▷ If the system is well designed and the mission well planned, the only remaining risk reduction strategies may be procedures and training.
  - ▷ Emergency procedures training and disaster preparedness exercises improve human response to hazardous situations.
- Create or improve the process.
- Create or improve checklists.
- Sequence of Events (Flow)—Put tough tasks first before fatigue, don't schedule several tough tasks in a row
- Timing (within tasks, between tasks)—Allow sufficient time to perform, to practice. Allow adequate time between tasks.
- Man-Machine Interface/Ergonomics—Assure equipment fits the people, and effective ergonomic design.
- Simplify Tasks—Provide job aids, reduce steps, provides tools like lifters communication aids.
- Reduce task loads—Set weight limits, automate mental calculations and some monitoring tasks.
- Back out options—Establish points where process reversal is possible when a hazard is detected.

Other Common Risk Control Options

Reject

- We can and should refuse to take a risk if the overall cost to a human life, equipment damage, or damage to the environment exceeds any benefits.
- We can and should refuse to take a risk if the overall costs of the risk exceed its mission benefits.
  - ▷ For example, a Part 141 pilot school may review the risks associated with formation flying. After assessing all the advantages of creating a new formation training option and evaluating the increased risk associated with it, even after application of all available risk controls, we may decide that the benefits do not outweigh the expected risk costs. UND Aerospace is better off in the long run by not training students to fly formation.

Avoid

- Avoiding risk altogether requires avoidance tactics.
- Examples and be changing the route, altitude, or destination.
  - ▷ You can avoid a crosswind by utilizing a different runway.
  - ▷ You can avoid the hazard of inflight icing by utilizing an aircraft certified for inflight icing.
  - ▷ You can avoid operations at a non-towered airport by landing earlier.
- Avoidance is an option that is often under-utilized to due to several human factors.

Delay

- It may be possible to delay a risk. If there is no time deadline or other operational benefit to speedy accomplishment of a risky task, then it is often desirable to delay the acceptance of risk.
- During the delay, the situation may change and the requirement to accept the risk may go away.
- Many environmental and fatigue risks can be successfully mitigated by a delay.

Transfer

- Risk transference does not change the probability or severity of the hazard, but it may decrease the probability or severity of the risk actually experienced by the individual or organization accomplishing the training flight/activity.
- As a minimum, the risk to the original individual or UND is greatly decreased or eliminated because the possible losses or costs are shifted to another entity.
- An example could be training a new student who is having difficulty landing the aircraft. Sometimes it makes sense to transfer the student to a new instructor who may have different communication skills.
- Transfer dangerous flight events, such as an emergency return to the field, to a flight simulator.

Spread

- Risk is commonly spread out by either increasing the exposure distance or by lengthening the time between exposure events.
- As an example, we can spread the number of airports used for touch and go landings.
  - ▷ Spreading 20 aircraft over 5 airfields is better than all 20 aircraft doing touch and goes at a single airport.
  - ▷ Encourage instrument approaches at a nearby satellite airport to lower the traffic count at Grand Forks International airport.
  - ▷ Encourage students to fly in the early morning before the traffic pattern becomes saturated.

Compensate (Create Redundancy)

- We can create redundant capability in certain special circumstances.
- An example is to plan for a backup. When a critical piece of equipment or other mission asset is damaged or destroyed we have capabilities available to bring on line to continue the mission.
  - ▷ Airlines always have spare crews on standby. If a pilot or flight attendant does not feel fit to fly a standby pilot or flight attendant can easily replace this person.
  - ▷ A spare aircraft may be used as a ground spare aircraft. If the primary aircraft has maintenance issue that may affect the safety of the passengers, the spare aircraft can be used.
  - ▷ Providing a spare power source for your iPad could allow a pilot to use their iPad, even after the battery goes dead.

Limit Exposure

- Establish work/rest cycles for high and low temperatures.
- Avoid takeoffs/landings during sunrise/sunset during bird migratory season.
- Limit the number of people or items exposed to a risk—during blizzard conditions expose only the essential personnel to the dangerous road conditions.
- Time—Minimize the time of exposure. Conduct the preflight in the hanger during very cold days. Tow the airplane out of the hanger after the crew is ready to start the engine(s).
- Iterations—Don't do it as often. Don't replace the light bulb at the top of the cell phone tower until the light bulb burns out.
- Turn off the engine cowling anti ice when flying in dry air.

Select Personnel

- Mental criteria—essential basic intelligence, and essential skills and proficiency
- Emotional criteria—essential stability and maturity
- Physical criteria—essential strength, motor skills, endurance, size
- Experience—demonstrated performance abilities

Rehabilitate

- Personnel—rehabilitation services restore confidence
- Facilities/equipment—get key elements back in service
- Mission Capabilities—alternate ways to continue the mission if primaries are lost

Motivate

- You can motivate through positive and negative incentives by establishing meaningful individual and group rewards or punishment.
- Another form of motivation is competition through the use of a healthy individual and group competition on a fair basis.
- You can also demonstrate the effects of unsafe acts through the use of graphic, dynamic, but tasteful demonstrations of the effects of unsafe acts.

Reduce Effects

The effects (consequences) of an unsafe act can often be reduced through the proper use of emergency equipment. Seat belts and air bags are good examples along with the use of fire extinguishers, first aid materials, spill containment materials.

STEP 6: MAKE CONTROL DECISIONS

Describe the measures you plan to use. Develop one or more controls for each hazard to reduce its risk.

STEP 7: DESCRIBE CONTROL IMPLEMENTATION

- Make sure to identify the personnel and organizations responsible for implementing the control(s).
- Examples could be:
  - ▷ The director of flight ops will write a new policy.
  - ▷ The director of aviation safety will coordinate and approve the new policy.
  - ▷ Publications will publish and control the new policy.
  - ▷ Flight leads will brief new policy at the next monthly flight ops meeting.
  - ▷ The director of maintenance and line will brief the new policy at their next meeting.
  - ▷ The director of aviation safety will create a new hot topics presentation and forward to the faculty.
  - ▷ The faculty will brief and discuss the new policy to the flight students.
- Remember to include the affected offices. Examples include:
  - ▷ Accountable Executive
  - ▷ Aerospace Foundation
  - ▷ Airport ATC
  - ▷ Airport Manager
  - ▷ Airport Ops
  - ▷ Airport TSA
  - ▷ Aviation Academic Chairperson
  - ▷ Aviation Safety Council
  - ▷ Avionics
  - ▷ Chief Flight Instructor (fixed-wing)
  - ▷ Chief Flight Instructor (helicopter)
  - ▷ Course Managers
  - ▷ Director of Aviation Safety (mandatory)
  - ▷ Director of Flight Ops
  - ▷ Director of Mx and Line
  - ▷ Dispatch
  - ▷ Fiscal Affairs
  - ▷ Flight Course Ground Instructors/ Professors
  - ▷ Flight Ops
  - ▷ Flight Ops Line
  - ▷ Flight Simulators
  - ▷ Helicopter Ops
  - ▷ Lead Instructors
  - ▷ Publications
  - ▷ Records
  - ▷ Scheduling
  - ▷ Standards
  - ▷ UAS

STEP 8: MEASURING NEW RISK CONTROLS

- Explain the approved implementation strategy for each control measure.
- How will we measure the success of these new risk controls? Measuring success is a part of our assessment process.
  - ▷ Examples may be to track online safety reports.
  - ▷ Flight data monitoring.
  - ▷ Spot inspections.
  - ▷ Employee surveys.
  - ▷ Student surveys.
  - ▷ Annual audits.
- A possible example could be:
  - ▷ The safety office will track the number of online safety reports—reports should decrease by 20 percent over 12 months.
  - ▷ The academics department will conduct a focus group survey in May—this will be a first ever report—after our first snapshot, following surveys should show a decrease below 25 percent.
  - ▷ The FDM analyst will measure exceedences beginning next month—rate of exceedences should decrease 10 percent.
  - ▷ Management will conduct and document spot inspections on a weekly basis—number of failed spot inspections should drop by 10 percent over 12 months.
  - ▷ The safety office will conduct spot inspections on a monthly basis—number of failed spot inspections should drop by 10 percent over 12 months and should show the same results as the management spot inspections.
  - ▷ Three questions will be added to the annual culture survey for students, staff, flight ops, and maintenance.
  - ▷ Add this topic to the internal auditor's checklist; next annual internal audit is January 2015.

STEP 9: RISK DECISION-MAKING

- In our example, the director of aviation safety will approve the authority and level for risk acceptance.
- The decisions for "Acceptable with Mitigation" risks will be properly coordinated with the appropriate organizations.
- Explain how we will ensure proper supervision before accepting this risk.
  - ▷ For example, the lead flight instructor authorizes the applicable CFIs and supervisors to approve a solo cross-country after the flight risk assessment tool is properly completed and the overall risk is "low" or "acceptable with mitigation."
- Make sure to identify the personnel and organizations responsible for implementing the control(s).

STEP 10: RISK ASSESSMENT AFTER RISK CONTROLS ARE IMPLEMENTED, REVIEWED, AND SUPERVISED

- Reassess and update the new likelihood and severity when all of the risk controls in blocks 3 through 9 are fully implemented and fully accomplished.
- Make risk decisions: decide to accept or reject the residual risk for this mission/task.
  - ▷ Attempt to lower the risk to a lower level: as low are reasonably practical.
- The acceptability of risk will be evaluated using the risk matrix. The risk matrix is color coded; unacceptable (red), acceptable (green), and acceptable with mitigation (yellow). The acceptance criteria and designation of authority and responsibility for risk management decision making is as follows (see also Figure 2-5 on page 26):
  - ▷ **Unacceptable (Red).** The risk is assessed as unacceptable and further work is required to design an intervention to eliminate the hazard or to control the factors that lead to higher risk likelihood or severity. The activity that causes exposure to this risk will be ceased at once, if it is ongoing. If the activity is not ongoing, it shall not be commenced, until further mitigation strategies can reduce the severity and/or likelihood of the event occurring to acceptable levels.
  - ▷ **Acceptable With Mitigation (Yellow).** Acceptable with mitigation by proper acceptance authorities. This risk may be accepted under defined conditions of mitigation. The acceptance will be by the appropriate director or dean, following mitigation as appropriate. In addition to the appropriate operational director, a director of safety must concur in the acceptance.
  - ▷ **Acceptable (Green).** Where the assessed risk falls into the green area immediate supervisors may accept it without further action. Safety risk management should always be used to reduce risk to as low as reasonable practicable (ALARP) regardless of whether or not the assessment shows that it can be accepted as is. This is a fundamental principle of continuous improvement
- The firector of aviation safety will determine the authority and level for risk acceptance.
- Decisions for unacceptable and acceptable with mitigation should be properly coordinated with the appropriate organizations.

STEP 11: SELECT A DATE FOR REVIEW

Typically this review will be conducted within 12 months. Some risk assessments may need to be reviewed much sooner.

STEP 12: REVIEW

- Indicate if the controls were effective or ineffective.
  - ▷ Did we achieve the measurable goals identified in Step 8?
  - ▷ Did we fail to anticipate any significant unintended consequences?

- For any ineffective controls, determine why and what to do the next time this hazard is identified. For example, change the control or change how the control will be implemented/supervised.

> **Note:** The completed worksheet will be sent to the Safety Office. The Safety Office retains all risk assessment worksheets.

TYING IT ALL TOGETHER: THE GENERAL PRINCIPLES OF SAFETY RISK MANAGEMENT

- Never take a risk that can be reasonably avoided.
- Acceptance of high risk tasks need to be approved by high level management.
- Risk assessments are accomplished before, during, and after the task.
- It is more important to establish clear risk mitigation procedures than to use generic approaches and procedures.
- There may be no "single solution" to a safety problem. There are usually a variety of directions to pursue.
- Each of these directions may produce varying degrees of risk reduction.
- A combination of approaches may provide the best solution.
- All system operations represent some degree of risk.
- Recognize that human interaction with elements of the system entails some element of risk.
- Keep hazards in proper perspective.
- Do not overreact to each identified risk, but make a conscious decision on how to deal with it (before, during, and after the task).
- Weigh the risks and make judgments according to your own knowledge, inputs from subject matter experts, experience, and program need.
- Subordinates need to communicate with supervisors/management concerning acceptable safety goals and how they can be achieved—not that their approach will not work.
- Supervisors/management officials need to communicate with their subordinates concerning acceptable safety goals and how they can be achieved—not that their approach will not work.
- There are no "safety problems" in task planning or design. There are only risk management problems that, if left unresolved, may lead to accidents.

CONCLUSION

You may implement the same SMS processes in different departments or areas of functional responsibility within your organization. For example, you may have an SRM process at the corporate level, at the maintenance division level, and within an engine repair department. This may be the same process (e.g., SRM) that has been implemented at the same carrier at three different levels of the organization: the corporate, division, and department.

You will need to continuously evaluate the manner in which the SRM process is implemented at each level. Do not assume that an SRM process implemented at the corporate level reflects equivalent implementation of this process at other levels. There is not an obligation for SMS to apply to non-aviation-safety-related processes.

REVIEW QUESTIONS

1. Explain how the SRM component of SMS provides a decision-making process for identifying hazards and mitigating risk

2. Analyze how the SRM concepts are based on a thorough understanding of your organization's systems and operating environment.

3. Describe how the SRM component is the organization's way of fulfilling its commitment to consider risk in their operations and to reduce it to an acceptable level.

CHAPTER 11
Safety Assurance and Continuous Monitoring

OBJECTIVES

- Understand how safety assurance measures the performance and effectiveness of your established risk controls.
- Learn how effective safety assurance activities identify new or emerging hazards.
- Describe the interaction of safety assurance and safety risk management.
- Explain the importance of objective evidence when tracking the performance of your Safety Management System.

KEY TERMS

- Aviation Safety Reporting System (ASRS)
- Continuous Monitoring Plan
- Non-Punitive Employee Safety Reports
- Objective Evidence
- Residual Risk
- Service Difficulty Reporting System (SDRS)
- Voluntary Safety Reporting Programs (VSRP)

INTRODUCTION

The third component of SMS is safety assurance (SA). In this component you are ensuring the performance and effectiveness of your established safety risk controls. These controls were established under the safety risk management process. Safety assurance is also designed to ensure that your organization meets or exceeds its safety objectives through the collection, analysis, and assessment of data concerning the company's performance.

The purpose of SA is for you to evaluate the overall effectiveness of risk controls and your safety management system. Your organization should be monitoring your systems and operations to ensure that new hazards are identified, ensure compliance with regulatory requirements applicable to the SMS, and ensure that your SA function is based upon the comprehensive system description as described in the system and task analysis process.

Your organization should collect the data necessary to demonstrate the effectiveness of your operational processes and the SMS. The outputs of SA activities feed other components of the SMS. For example, when SA activities identify hazards, we are driven back to the SRM process where we again analyze and assess the associated risk and design new risk controls as necessary. This is done through a continuing process of data collecting methods such as audits, evaluations, investigations, employee reports and continuous monitoring processes. The data collected is analyzed and then converted into information. This information is used to verify the performance and effectiveness of risk controls, identify new hazards and assure your organization's compliance with the Code of Federal Regulations.

Safety assurance also provides confidence that, over time, risk controls remain relevant and effective as your organization changes and that safety practices are continuously improved upon through structured management reviews.

Safety assurance works in close partnership with SRM as well as feeds the other components of your safety management system. The SA process parallels quality assurance by ensuring discrepancies are documented, solutions implemented, and corrective actions recorded. However, safety assurance ensures that safety objectives are being met and, in this regard, differs from quality assurance. If your organization currently has quality assurance processes/programs and company personnel and resources dedicated to those functions, they can/should be leveraged to support your SA requirements.

Safety assurance might be defined as activities that are designed to gain confidence that the risk controls established during the SRM continue to be effective. The SA function applies the quality assurance and internal evaluation activities to ensure that risk controls continue to conform to their expectations and that they continue to be effective in maintaining a risk within acceptable levels. These assurance and evaluation functions provide a basis for continual improvement.

Safety assurance consists of the following seven data acquisition processes:

1. Monitoring of operational processes;
2. Monitoring of the operational environment;
3. Auditing of operational processes;
4. Evaluation of the SMS and operational processes;
5. Investigations of accidents and incidents;
6. Investigations of potential noncompliance; and a
7. Confidential employee reporting system.

SA AND INTERACTIONS WITH SRM

As shown in Figure 2-7 on page 38, in SA the process continues with measuring and monitoring the performance of the system operation and system monitoring, with the designed risk controls in place. This involves a variety of data sources (data acquisition) that will be further explained in this chapter. As in SRM, the data will need to be analyzed in order for it to be used in decision-making (analysis of data). In the case of SA, the decision-making can result in several paths (system assessment). If the data and analysis say that the system and its risk controls are functioning as intended, the result is confirmatory: the management now can have confidence in system safety performance.

If this is not the case, the analysis needs to continue to determine if the shortfall is due to the fact that the controls are not being used as intended, or if, even though the system is being used as intended, it is not producing the expected results. In the former case, action should be taken to correct the problem (corrective action). In the latter case, the system design should be reconsidered using the path back to the SRM process.

The path back to SRM is a particularly important part of the SA process, especially for organizations that are transitioning into SMS. Their operational systems have likely not been built using an SRM process, so they may lack formal or well-understood risk controls. The SA process covers the day-to-day life of system operations; so the determination to review existing processes for hazard and risk may be the first time that these aspects of operation have been considered.

As in SRM, managers who are responsible for operational processes are the ones who are also responsible for assuring that they are performing as intended from a safety and operational standpoint. Moreover, correct design, performance, and risk controls need to be a concern of top management, including the accountable executive.

REQUIRED CONTINUOUS MONITORING

After the completion of a safety risk assessment, the monitoring plan should be comprehensive to verify the predicted residual risk. A **continuous monitoring plan** includes the safety performance targets or another sound method to verify the predicted residual risk. Create a plan for each hazard that defines:

- Monitoring activities;
- The frequency and duration of tracking monitoring results; and
- How to determine, measure, and analyze any adverse effects on adjoining systems.

Monitoring Activities

The monitoring organization must verify that the existing controls and/or safety requirements were indeed put in place and are functioning as designed. Specifically, this means that procedures must be stringently followed and hardware or software must function within the established design limits.

Detail the methods by which the risk acceptor's designee will gather the performance data or monitoring results. The organization that accepted the risk is responsible for comparing the monitoring results against the defined safety performance targets or using the results to determine whether predicted residual risk was met.

It is important to retain objective evidence that the mitigations have been implemented. Objective evidence is simply documented proof. The evidence must not be circumstantial; it must be obtained through observation, measurement, testing, or other means.

Frequency and Duration of Monitoring

When considering the frequency and duration of tracking monitoring results, account for:
- The complexity of the change;
- The hazard's initial risk level;
- How often the hazard's effect is expected to occur (i.e., likelihood);
- Existing controls;
- The types of safety requirements that are being implemented (if any); and
- The amount of time needed to verify the predicted residual risk.

For example, when considering a hazard associated with the familiarity of a new procedure, a relatively short tracking period would be required until a person or population could reasonably be expected to adapt to the new procedure and the predicted residual risk could be verified. The monitoring plan for a hazard associated with new separation criteria, however, may require several years of tracking to verify the predicted residual risk.

Documenting the Monitoring Plan

Table 11-1 on the next page provides an example of a monitoring plan. As a minimum, the monitoring plan should include the following information for each hazard:
- Hazard;
- Hazard's initial risk;
- Hazard's predicted residual risk;
- Existing controls and/or safety requirements;
- Monitoring activities and their frequency and duration;
- The organization responsible for implementing safety requirements; and
- Safety performance targets.

Continuous monitoring to ensure continuous improvement. It is critical to obtain feedback on safety performance indicators through continuous monitoring. Organizations responsible for performing quality control and/or quality assurance use audits and assessments to monitor the safety risk and performance of an implemented change documented in the monitoring plan. The responsible organization determines whether an implemented change is meeting the safety performance targets documented in the monitoring plan.

Results of post-implementation monitoring help determine whether a change can be made permanent or must be reassessed through the SRM process.

Monitoring Current Risk

A hazard's current risk is updated at each monitoring interval (in accordance with stated monitoring frequency). Current risk provides an indicator of whether safety requirements are meeting the predicted residual risk. The risk acceptor assesses the current risk as often as prescribed for the duration of the monitoring plan.

(1) Hazard ID	(2) Hazard Description		(3) Initial Risk
ABC-M01	Aircraft access unauthorized surface		High (3B)
(4) Safety Requirements		(5) Organizational Responsibility for Implementation	
1. Paint two new markings to indicate the specific locations of runway 8 and 12 near the intersection of runway 8 and taxiway E-1. 2. Provide NOTAMs. 3. Update AFD note to include the use of non-standard placement marking. 4. Update hot spot directory. 5. Brief pilots. 6. Provide letters to airmen.		1-3: Joe Smith, ABC District office 4: Jane Smith, Central Service Center 5: Jerry McGuire, ABC Operations Manager 6: Ken Anderson, ABC Flight Standards District Office	
(6) Predicted Residual Risk	(7) Monitoring Activities		
Medium (3C)	Maintain a daily log detailing any event in which an aircraft aligns on the incorrect runway. Measure the number of pilot deviations assoicated with pilots unaware of the NOTAMs, AFD note, hot spot data, pliot briefings, and letters to airmen. Safety requirement 1 will be verified once for implementation and compliance with code. Safety requirements 2-6 will be monitored to ensure implementation is completed and publications are updated and issued.		
(8) Frequency	(9) Duration		(10) Safety Performance Targets
Review the collected daily logs monitoring any aircraft aligning on the incorrect runway and the collected questions and concerns from the pilot community on a monthly basis.	For a minimum of one year, or until the performance targets are met and the predicted residual risk is verified.		Fewer than two surface misalignments within a year (estimated 300,000 operations annually).

Table 11-1   Example of a monitoring plan.

What if the predicted residual risk is not achieved?

Through monitoring current risk and the safety performance of a recently implemented change, it may become clear that the predicted residual risk is not being met. If this occurs, the safety analysis must be revisited to assess the risk of the new hazards or develop additional safety requirements to lower the risk to an acceptable level. There are several reasons why this may occur:

- The safety requirements or existing controls may not be properly mitigating the risk;
- The initial risk may have been assessed inaccurately;
- Unintended consequences occurred; or
- New hazards are identified.

In either case, the risk acceptor must coordinate a reassessment to determine if changes to the mitigation strategy are necessary. An SRM panel must be convened to assess the risk of the new hazards and/or develop additional safety requirements to lower the risk to an acceptable level.

Predicted Residual Risk Met

The successful completion of monitoring is a prerequisite to hazard and NAS change closeout. This includes the achievement of safety performance targets and/or the predicted residual risk.

The monitoring procedures used to verify the predicted residual risk must also be documented, as they will be used to evaluate the safety performance of the change after it is added to the organizations policies or procedures. The established monitoring requirements must be followed, even after meeting the goals of the monitoring plan.

> **Residual risk** is the level of risk that has been verified by completing a thorough monitoring plan with achieved measurable safety performance target(s). Residual risk is the assessed severity of a hazard's effects and the frequency of the effect's occurrence.

Monitoring and Tracking of Changes to Your Processes

A change is considered to be part of the new operating process only after monitoring is completed, the safety performance target is achieved and maintained, and/or the predicted residual risk is verified. At that point, the change is monitored through existing safety assurance processes to determine whether an acceptable level of safety is maintained. The change and all of the associated safety requirements become part of the operating process, which will become the basis from which all future changes will be measured. If a safety requirement is altered or removed from a newly created process, a new SRM analysis must be performed.

The documentation that was developed during the SRM process is critical to SA functions, which often use SRM documents as inputs to assessments and evaluations.

DATA ACQUISITION

The outputs of safety assurance activities feed other components of the SMS. This is done through a continuing process of data collecting methods such as continuous monitoring processes, employee reports, investigations, audits, and evaluations. The data collected is analyzed and then converted into information. This information is used to verify the performance and effectiveness of risk controls, identify new hazards and assure your organization's compliance with the Federal Regulations. A few examples of continuous monitoring mechanisms follow.

FLIGHT OPERATIONAL QUALITY ASSURANCE (FOQA). FOQA is a voluntary safety program designed to improve aviation safety through the proactive use of flight-recorded data. Operators use the data to identify and correct deficiencies in all areas of flight operations. Properly used, FOQA data can help carriers take action to reduce or eliminate safety risk, as well as minimize deviations from regulations. Through access to de-identified aggregate FOQA data, the FAA can identify and analyze national trends and target resources to reduce operational risk in the NAS, ATC, flight operations, and airport operations.

AVIATION SAFETY ACTION PROGRAM (ASAP). The purpose of ASAP is to prevent accidents and incidents by encouraging certificate holder employees to voluntarily report safety issues and events. ASAP provides for the education of appropriate parties and the analysis and correction of safety concerns that are identified in the program. ASAPs are intended to create a nonthreatening environment that encourages employees to voluntarily report safety issues even though they may involve violation of 49 USC, Subtitle VII, or violation of 14 CFR. ASAP is based on a safety partnership between the FAA and the certificate holder and may include any third party, such as an employee labor organization. ASAP allows the reporting and collecting of safety information that may not otherwise be obtainable.

Through the analysis of ASAP data, potential precursors to accidents can be identified. The FAA has determined that identifying these precursors is essential to further increasing aviation safety. Under an ASAP, safety issues are resolved through corrective action rather than through punishment or discipline, and it can help to educate appropriate parties in preventing a reoccurrence of the same type of safety event.

VOLUNTARY SAFETY REPORTING PROGRAMS (VSRP). The Air Traffic Safety Oversight Service (AOV) provides guidance for establishing an ASAP for Air Traffic Organization (ATO)-credentialed safety personnel in SOC 07 – 04, Aviation Safety Action Program (ASAP) For Credentialed ATO Personnel. The objective of the program is to encourage credentialed personnel to voluntarily report safety information that may be critical to identifying potential precursors to accidents. Under this guidance, safety-related issues are resolved through corrective action rather than through punishment or discipline. The ATO, in cooperation with its employee labor organizations and AOV, established two voluntary safety-reporting programs for controllers and technicians called the Air Traffic Safety Action Program (ATSAP) and the Technical Operations Safety Action Program (T-SAP).

The ATO VSRP is modeled after the successful ASAP program used in the aviation industry. Specifics on the ATO VSRP are contained in JO 7200.20, Voluntary Safety Reporting Programs. ATO employees voluntarily identify and report safety and operational concerns. The collected information is reviewed and analyzed to facilitate early detection and improved awareness of operational deficiencies and adverse trends. The information specified in employee reports is used to identify root causes and determine appropriate remedial actions, which are then monitored for effectiveness. This process promotes collaboration between employee work groups and management for the early identification of hazards and to maintain a proactive approach regarding safety concerns and corrective action recommendations.

AVIATION SAFETY REPORTING SYSTEM (ASRS). The ASRS is an important facet of the continuing effort by government, industry, and individuals to maintain and improve aviation safety. ASRS collects voluntarily submitted aviation safety incident/situation reports from pilots, controllers, and others. It acts on the information these reports contain and identifies system deficiencies, and issues, alerting messages to persons in a position to correct them. The ASRS educates through its newsletter *CALLBACK*, its journal *ASRS Directline*, and through its research studies. Its database is a public repository that serves the needs of the FAA, NASA, and other organizations worldwide that are engaged in research and the promotion of safe flight.

ASRS collects, analyzes, and responds to voluntarily submitted aviation safety incident reports in order to lessen the likelihood of aviation accidents. ASRS data is used to:

- Identify deficiencies and discrepancies in the NAS so that these can be remedied by appropriate authorities;
- Support policy formulation and planning for, and improvements to, the NAS; and
- Strengthen the foundation of aviation human factors safety research. This is particularly important since it is generally conceded that over 66 percent of all aviation accidents and incidents have their roots in human performance errors.

SERVICE DIFFICULTY REPORTING SYSTEM (SDRS). The SDRS is another reporting system in which aircraft owners/operator can report, via a web-based system, maintenance and/or service problems for any aircraft, engine, or component. SDRS is mandatory for commercial operators only. Certificate holders are required to report the occurrence or detection of each failure, malfunction, or defect.

NON-PUNITIVE EMPLOYEE SAFETY REPORTS. All companies should strive to develop a non-punitive, disciplinary policy as part of their SMS. Employees are more likely to report events, and cooperate in an investigation, when some level of immunity from disciplinary action is offered. When considering the application of a non-punitive disciplinary policy, companies might want to consider whether the event involved willful intent on the part of the individual involved, and the attendant circumstances. For example, has the individual been involved in an event like this before, and did the individual participate fully in the investigation.

A typical non-punitive reporting policy might include the following statements:

- Safe flight and ground operations are *[insert your organization]* most important commitment. To ensure this commitment, it is imperative that we have uninhibited reporting of all incidents and occurrences that compromise the safety of our operations.
- We ask that each employee accept the responsibility to communicate any information that may affect the integrity of flight and ground safety. Employees must be assured that this communication will never result in reprisal, thus allowing a timely, uninhibited flow of information to occur.
- All employees are advised that *[insert your organization]* will not initiate disciplinary action against an employee who discloses an incident or occurrence involving flight or ground safety. This policy cannot apply to criminal, international or regulatory infractions.
- *[Insert your organization]* has developed safety reports to be used by all employees for reporting information concerning flight safety or ground safety. They are designed to protect the identity of the employee who provides information. These forms are readily available in your work area.
- We urge all employees to use this program to help *[insert your organization]* continue its leadership in providing our customers and employees with the highest level of flight and ground safety.

Such a policy should be clearly laid out and communicated to all staff. Some operators communicate this policy to their staff by having it printed on the hazard reporting forms. In order to encourage a healthy reporting culture in a company, there should really be only three reasons to discipline an employee. They are:

- Willful negligence;
- Criminal intent; and
- Use of illicit substances.

CONCLUSION

Using the principles of safety assurance your organization should collect the data necessary to demonstrate the effectiveness of your operational processes and the SMS. The outputs of safety assurance activities feed other components of the SMS. For example, this is how you create objective evidence that you are doing what you say you are doing. Safety assurance also provides confidence that, over time, risk controls remain relevant and effective as your organization changes and that safety practices are continuously improved upon through structured management reviews.

REVIEW QUESTIONS

1. Explain how SA measures the performance and effectiveness of your established risk controls.

2. Explain effective SA activities used to identify new or emerging hazards.

3. Describe the interaction of SA and safety risk management.

4. Explain the importance of objective evidence when tracking the performance of your SMS.

CHAPTER 12
Safety Assurance and Audits

OBJECTIVES

- Understand the need for audits and evaluations.
- Know the difference between an audit and an evaluation.
- Know the difference between an internal and external audit/evaluation.
- Know the purpose of a preliminary gap analysis.
- Know the purpose of a detailed gap analysis.
- Know the definition of objective evidence.
- Know how to conduct an audit and evaluation using the FAA's SMS Voluntary Program Guide.

KEY TERMS

- Confirmation
- External Audit/Evaluations
- Internal Audit/Evaluations
- Objective Evidence
- Physical Examination
- Tracing
- Vouching

INTRODUCTION

We will continue the discussion of safety assurance, focused on the two remaining processes used for data acquisition: auditing of operational processes along with the evaluation of the SMS and operational processes. These audits and evaluations can be accomplished either internally or externally. We will refer to audits/evaluations conducted within your organization as an internal audit/evaluation. Audits or evaluations conducted by personnel outside your organization will be referred as an external audit or evaluation. The highest level of external evaluation is conducted by the FAA. Alternatively, the external evaluation can be performed by another agency that recognizes your SMS Program.

AUDITING TOOLS

Several auditing and inspection tools can be used for assessing your SMS. These tools are used for initial assessment or ongoing surveillance and oversight. The auditing and inspection tools are based on a series of indicators that help an auditor (internal or external) assess the effectiveness of an organization's SMS. These require an interaction with the organization including face-to-face discussions and interviews with a cross-section of people as part of the assessment. Auditing recognizes the difference in oversight methodologies from traditional compliance-based oversight to performance-based oversight that assesses not only compliance but also the effectiveness of the SMS. It has been designed to indicate the expected standard of an organization's SMS in terms of compliance with the SMS regulation and its performance to effectively manage safety risk. It has been developed to harmonize and standardize an approach to SMS globally establishing an equivalent standard of SMS oversight. Furthermore, these tools have been designed to allow virtually any organization to use and adapt these tools to serve its own purposes (rather than developing a tool from nothing).

An initial step in developing an FAA recognized SMS to conduct a gap analysis in order to analyze and assess your existing programs, systems, processes, and activities with respect to the FAA SMS functional expectations found in the SMS framework and the SMSVP.

Depending upon the size and complexity of your organization, the detailed gap analysis may take four to six months to complete. The detailed gap analysis is a living process and will be continuously updated as SMS implementation progresses. Completing a gap analysis allows you to determine what existing programs, processes, and practices comply with the SMSVP standard and identify those that do not.

Initial Assessment Tools

The gap analysis tool can be used as part of an initial assessment and should define the expectations on the individual indicators for your SMs journey. For example, an initial assessment could be based on a desk top review of the documentation that focuses on assessing whether the "indicators for compliance and performance" are present and suitable. The grading criteria are shown in Table 12-1. Obviously, your preliminary gap analysis will have many items graded as "Not Performed" or "Planned." The number of criteria already being performed by your organization might surprise you!

Assessment Level	Assessment Rating Scale Word Picture	Assessment Scale
Not Performed	No action has been taken on this requirement.	NP
Planned	A plan exists with actions to be taken or manual(s) affected, a scheduled completion date and responsible individual or group identified to meet this requirement.	PLN
Documented*	The policy and procedural guidance of this requirement are incorporated into company documents such as manuals, training materials, and work instructions.	DOC
Accomplished*	Resources are in place to accomplish all objectives of this requirement. Employees are expected to be trained and knowledgeable on the policies and procedures which were documented in the previous assessment level (DOC).	ACC
Demonstrated*	This requirement of your company's SMS has been subjected to at least one round of evaluation/ auditing by your company (and validated by the CMT/CHDO) to demonstrate operational performance.	DEM

These assessments require organization (FAA Certificate Management Team, FAA SMS Program Office, etc.) oversight validation.

Table 12-1   Grading criteria for preliminary and detailed gap analysis.

Once the desktop review has been satisfied an on-site visit should be carried out to assess whether the indicators are operating and overall effectiveness is achieved. The on-site visit should normally be carried out by a team including a team leader with an appropriate level of competence in SMS and technical specialists to support the assessment.

It is important to structure the assessment in a way that allows interaction with a number of people at different levels of the organization to determine how effective aspects are deployed throughout the organization. For example, to determine the extent that the safety policy has been promulgated and understood by staff throughout the organization will require interaction with a cross-section of staff. For small organizations it may be more practical to have a single assessor appropriately trained in SMS and with the technical competencies to assess the organization.

Another approach is for the regulated organization to partially complete the tool as a self-assessment, including the "how it is achieved" box, and submit this to the regulator, who would decide whether it was sufficiently progressed to warrant an on-site visit and then verify and validate the organizations self-assessment.

Ongoing Surveillance Tools

For ongoing surveillance auditors may also define expectations for individual indicators. However all individual "indicators of compliance and performance" should be at least operating and that effectiveness is achieved in all of the elements.

Auditor Competencies, Training, and Qualifications

If feasible, your organization should specify that your auditors have training and/or experience in recognized quality management auditing, systems analysis, root cause analysis, and risk assessment, as well as evaluation principles and techniques. The assessment tools used during an audit or evaluation should be used by regulatory staff with training and competency in:

- SMS based on the FAA and ICAO SMS framework;
- Understanding of Quality Management Systems, compliance and auditing;
- Interview techniques;
- Understanding of risk management;
- Appreciation of the difference between compliance and performance;
- Report writing techniques to allow narrative to be used to summarize the assessment;
- In-house prepared courses;
- College courses;
- Home study course materials;
- Industry seminars and workshops; and
- Selected FAA Courses.

It is recommended that as well as being trained to use the tool in the classroom environment, staff are provided additional training during a live assessment to familiarize themselves with the tool and its practical use.

Instructions for Using the Gap Analysis Tool

The FAA gap analysis tool evaluates the compliance and effectiveness of the SMS through a series of indicators. It is set out using the elements of the ICAO SMS framework with the framework definition followed by an effectiveness statement for that element. For each element, a series of indicators for compliance and performance is listed followed by a series of indicators of best practice. Each indicator should be reviewed to determine whether the indicator is present, suitable and operating and effective, using the definitions and guidance set out below, so that the overall effectiveness of the element can be justified and supported. The tool would normally be used by the regulator to record and document the assessment. Alternatively, it can be partially completed by the organization to assess itself ("How it is achieved" column) and by the regulator to verify and validate the organization's assessment ("Verification" column and "Summary comments" box). Applicability The evaluation tool can be used to assess any regulated organization. Due consideration should be given to the size, nature and complexity of an organization in carrying out the assessment. For smaller organizations a reduced number of indicators may be used as defined by the regulator.

Evidence

Evidence includes documentation, reports, records of interviews and discussions and is likely to vary for different levels of indicator assessment. For example, for an indicator to be present the evidence is likely to be documented only, whereas for assessing whether it is operating it may involve assessing records as well as face to face discussions with personnel within an organization. "How it is achieved" should include summary statements and any references to documentation and records. The Verification Column should be for the regulator to record any observations, conversations, records and documents sampled.

Summary Comments

Once the auditor has assessed all indicators, a judgement can be made on whether the overall effectiveness of the evaluation element has been achieved; this should be noted in the summary comments box.

Modifying the Tools

An organization may adapt the terminology and tool to meet its own specific requirements. The FAA has attempted to make all of the horizontal rows applicable to most aviation organizations interested in implementing an FAA Recognized SMS Program. If a specific row of the spreadsheet is out of scope of your system, simply mark the row as "Not Applicable". The vertical columns need to be modified to support your specific organization. As an example, a Part 141 pilot school may modify the vertical columns to be: Safety Department, Flight Operations, Maintenance, and Academics. The overall grade for any single row is the lowest grade of the four departments.

Developing Procedures

Each regulator will need to define procedures around the use of the tool, customized to its own organizational structure and approach to SMS oversight activity.

Complementary Assessment Tools

Multiple tools will be used throughout your SMS journey. For most organizations SMS will take time to implement and several years to mature to a level where it is effective. The typical saying is, "You have to crawl before you walk, and you have to walk before you run." You will see increasing levels of SMS maturity as your organization implements and develops its SMS. Hence your tools need flexibility in order to fairly assess the indicators against the organization's SMS maturity. The evaluation tools can be used in stages looking initially for whether the key elements of an SMS are present and suitable. At a later stage the SMS can be assessed for how well it is operating and effective but it also recognized best practice. Your organization can always strive towards excellence as part of their continuous improvement programs and the tool allows that best practice to be assessed.

WHAT DOES A MONITORING PROGRAM INVOLVE?

A quality evaluation and auditing program involves some basic methods and procedures common to many forms of management reviews. Below are nine common practices used in program monitoring (for safety or quality).

- **Physical examination (PE).** The activity of gathering physical evidence. This is a substantive test involving the counting, inspecting, gathering, and inventorying of physical and tangible assets such as cash, plants, equipment, parameters, etc.
- **Confirmation.** The act of using a written response from a third party to confirm the integrity of a specific item or assertion.
- **Vouching.** The examination of documents that support a recorded transaction, parameter, or amount. Testing starts with the recorded item and moves on to review the supporting documentation.
- **Tracing.** The following of source documents to their recording in the accounting records. This is a "through the system" method of accounting transaction flows, ledgering accounts, or logging parameters.
- **Performance.** An auditing technique of repeating a client process or activity with high fidelity and comparing results with previous operational data.
- **Observation.** The process of witnessing physical activities of the client. It differs from the PE in that the auditor observes the client performing the client's process rather than the auditor performing the examination.
- **Reconciliation.** The process of matching two independent sets of records. The key here is independent. A derived set of data from the client does not meet this criterion, only third party or certified independent data meets the criteria. This satisfies the test of completeness and existence of evidence.
- **Inquiry.** The technique of asking questions and recording the response.
- **Inspection.** The critical examination of documents (different from vouching or tracing) to determine content and quality of a transaction such as inspecting leases, contracts, minutes of meetings, requirements, or client policy.

The operational audit addresses the effectiveness and efficiency of the organization. The objective is to determine the organization's ability to achieve its goals, objectives, and mission.

The compliance audit is an evaluation or assessment of conformance to established criteria, process, or work practice. The objective is to determine if employees and processes have followed those policies and procedures established by management.

WHAT IS THE DIFFERENCE BETWEEN AN EVALUATION AND AN AUDIT?

Evaluations are performed on site, while audits are an offsite method for assessing the facility.

An audit is a methodical, planned review used to determine how business is being conducted and compares results with how business should have been conducted in accordance with established SMS procedures. Audits are accomplished through discussions with a small number of personnel, and/or review of requested data, manuals, meeting minutes, records, and documentation. The typical term is objective evidence.

Evaluations are accomplished through direct observation, data monitoring, attendance at personnel meetings, observation of training activities, review of administrative and maintenance records and reports, and interviews or discussions. Interviews involve many of the following: accountable executive, process owners, process managers, supervisors, support and technical specialists, union representatives, employee participation group representatives, and other facility personnel. An evaluation is a functionally independent review of company policies, procedures, and systems. If accomplished as an internal evaluation, the evaluation should be done by an element of the company other than the one performing the function being evaluated. The evaluation process builds on the concepts of audit and inspection. An evaluation is an anticipatory process, and is designed to identify and correct potential findings before they occur. An evaluation is synonymous with the term systems audit.

Through audits and evaluations the overall effectiveness of the SMS is evaluated. As described in Chapter 11, safety data is tracked and analyzed for adverse trends and to identify the need for safety enhancing measures. Since the goal of the SMS is to increase the safety of your organization by meeting or exceeding safety objectives, the SMS is evaluated on your ability to manage the safety risk in your organization and meet these objectives. The result of an audit and evaluation will:

- Review and provide input on safety risk assessments;
- Review and provide recommendations regarding safety risk management processes and SRMDs;
- Review and provide input on the results of safety assurance functions within FAA organization;
- Review safety data analysis reports; and
- Analyze safety data and advise senior management on safety related issues.

SMSVP GUIDE AND JOB AIDS

With the newly created 14 CFR Part 5, the FAA published AFS-900-002-G201, SMS Voluntary Program Guide. This guidance describes the methods required of Flight Standards certificate management offices to assess an aviation service provider's formal, voluntary application of SMS requirements.

AFS-900 is the office of primary responsibility for the Flight Standards SMS Voluntary Program. The SMS Voluntary Program is available to all internal FAA Flight Programs Offices and all FAA certificated, external aviation service providers. The SMSVP meets State recognition requirements as defined in ICAO, Annex 6, Operation of Aircraft Part I, International Commercial Air Transport—Aeroplanes.

The SMSVP is how the FAA conforms to the International Civil Aviation Organization (ICAO) definition of an SMS "acceptable to the State". An SMS required by regulation or developed within a voluntary program corresponds to ICAO SMS requirements and will be accepted by other ICAO Member States.

> **Note:** Certificate holders in the SMSVP must meet program requirements for FAA recognition.

The SMSVP Design Job Aids

The Design Job Aids are used to evaluate your organization's documentation (objective evidence) describing its SMS applications. The completed job aids will become the FAA inspector's formal record of observations and evaluations.

1.3 - Designation & Responsibility of Required Safety Management Personnel		
1)	Does the certificate holder's processes require that all members of management develop, implement and maintain SMS processes within their area of responsibility to include, but not limited to: · Hazard identification and safety risk assessment; · Assuring the effectiveness of safety risk controls; · Promoting safety as required in subpart E, Safety Promotion; and · Advising the accountable executive on the performance of the SMS and on any need for improvement?	☐ Yes ☐ No
SMS Part 5 Rule 03-09-2015: 5.23(a)(2)		
Remarks:		
2)	Do the certificate holder's safety management processes identify the levels of management with the authority to make decisions regarding safety risk acceptance?	☐ Yes ☐ No
SMSVP Standard 03-09-2015: 5.23(b)		

Figure 12-1 Example SMSVP Job Aid. *(FAA)*

The FAA Certificate Management Team will validate, to the extent possible, that your organization's process design conforms to the SMSVP Standard. The Design Job Aids encompass the operational SMSVP conformance requirements. These job aids are considered the minimum performance validation activities to be used during the design validation phase.

Job Aid References:

- SMSVP Design Validation Job Aid—Policy, Attachment 2
- SMSVP Design Validation Job Aid—Safety Risk Management (SRM), Attachment 3
- SMSVP Design Validation Job Aid—Safety Assurance, Attachment 4
- SMSVP Design Validation Job Aid—Safety Promotion, Attachment 5
- SMSVP Performance Demonstration—Safety Policy, Attachment 6
- SMSVP Performance Demonstration—Emergency Preparedness and Response Process, Attachment 7
- SMSVP Performance Demonstration—Process Owner/Department SRM Process, Attachment 8
- SMSVP Performance Demonstration—Corporate SRM Process, Attachment 9
- SMSVP Performance Demonstration—Audit Process, Attachment 10
- SMSVP Performance Demonstration—Evaluations Process, Attachment 11

- SMSVP Performance Demonstration—Investigations Process, Attachment 12
- SMSVP Performance Demonstration—Continuous Improvement Process Attachment 13
- SMSVP Performance Demonstration Test—Accountable Executive Review Process, Attachment 14
- SMSVP Performance Demonstration—Records Retention Process, Attachment 15
- SMSVP Performance Demonstration—Safety Communications, Attachment 16

CONCLUSION

The process of audits and evaluations include assessment of all operational functions of your SMS. Audits and evaluations, both internal and external, are fundamental elements of the overall management system that includes organizational management, documentation, safety programs, quality assurance, and emergency response planning. The thrust of such initiatives is to effectively integrate these functions into the management system.

Development of audits and evaluations, as discussed in this chapter, should ensure that or organization's SMS policies and procedures are responsive to organizational changes and that certificate holders continually comply with appropriate safety and regulatory requirements. Furthermore, the FAA strongly encourages certificate holders to make an audits and evaluations an integral part of their management process.

REVIEW QUESTIONS

1. What is the difference between an audit and an evaluation?

2. What is the difference between an internal audit/evaluation and an external audit/evaluation?

3. Explain the purpose of a preliminary gap analysis.

4. Explain the difference between and preliminary gap analysis and a detailed gap analysis.

5. Define the term objective evidence.

6. Explain how to use the Job Aids in the FAA SMS Voluntary Program Guide.

CHAPTER 13
SMS and Your Safety Culture

OBJECTIVES

- Develop an understanding of what a culture is.
- Analyze components that make up a positive culture in regard to safety.
- Evaluate the implications of having a "non-punitive" or "just" culture within an organization.
- Identify methods to improve the safety culture within your organization.
- Develop an understanding of how to measure your safety culture at your organization.

KEY TERMS

- Closed Culture
- Confidential Employee Reporting System
- Safety Culture

INTRODUCTION

Safety culture is of central importance to having an SMS. In fact, it acts as the wrapper that holds the four SMS components together. You may have the best written safety policy, the most skillful developed risk matrix chart, and the most robust audit system, along with a communication network like no other, but if you do not have a positive organizational culture in relation to safety, it all means very little. Throughout this chapter we will explore what a positive safety culture is, how to recognize it, analyze the various components involved, and provide practical examples that you can use to begin development of a more positive culture within your organization.

DEFINING SAFETY CULTURE

There are many definitions for safety culture. Some argue that it can neither exist nor can it be measured. Experientially and through significant research, it is our opinion that it can be defined, measured, and with an experience safety professional it can quickly recognized within an organization. The old cliché, "if it looks like a duck, sounds like a duck, walks like a duck, and smells like a duck, then it's probably a duck" fits well with a safety culture. Within an organization there are signs all over that represent the general culture of the organization. It may be the dress code, the airport badge around everyone's neck, what is celebrated as success, and how your employees are evaluated and rewarded. Policies, procedures, and processes within your organization start to form a framework for what is truly considered important. Trained eyes to the visual cues, ears to hear what is being said compared to what is written, and some experience from safety professionals who have been exposed to different organizations, can start to identify clear sign posts that look into the center of an organization and define and measure its culture.

Culture is a general term, something we have been taught about since history or geography class in high school. It is a "way of living" for your organization. Safety culture is our behaviors and behavior characteristics within the context of trying to reduce and management risk and measuring that effort, while allowing your organization to accomplishing its mission.

SAFETY MANAGEMENT AND SAFETY CULTURE

One key aspect that is essential to safety performance is the culture of the organization. **Safety culture** is the term that we apply to those aspects of the organization's culture that relate to safety performance. The concept of safety culture underlies safety management and is the basis for the SMS requirements of 14 CFR Part 5 and the SMSVP Guide.

Because the culture of an organization includes the deeply ingrained and automatic psychological and behavioral aspect of human performance, there is a strong correlation between safety culture and accident prevention. Therefore, safety culture and SMS are interdependent. Management's constant attention, commitment, and visible leadership are essential to guiding an organization toward a positive safety performance.

The NTSB, which investigates accident and incident, has clearly made this connection and research shows that it is becoming a significant part of the discussion as we move away from blaming the who and start being more concerned with the organizational questions that ask what, how, and why. In NTSB Accident Report NTSB/AAR-14/03 PB2014-108877 titled "Crash Following Encounter with Instrument Meteorological Conditions After Departure from Remote Landing Site Alaska Department of Public Safety Eurocopter AS350 B3, N911AA Talkeetna, Alaska March 30, 2013" states on page one, "Safety issues include inadequate pilot decision-making and risk management; lack of organizational policies and procedures to ensure proper risk management; inadequate pilot training, particularly for night vision goggle use and inadvertent instrument meteorological condition encounters; inadequate dispatch and flight following; lack of a tactical flight officer program; punitive safety culture; lack of management support for safety programs; and attitude indicator limitations."

The NTSB states in the Executive Summary page iv, "An organizational safety culture that encourages the adoption of an overly punitive approach to investigating safety-related events tends to discourage the open sharing of safety-related information and to degrade the organization's ability to adapt to operational risks." Further emphasizing the point, page VIII continued, "The investigation found that Alaska DPS had a punitive safety culture that impeded the free flow of safety-related information and impaired the organization's ability to address underlying safety deficiencies relevant to this accident." The NTSB and the FAA have made is abundantly clear safety culture is real and management's role over time is critical.

Cultures are the product of the values and actions of the organization's leadership as well as the results of organizational learning. Cultures are not really "created" or "implemented;" they emerge over time and as a result of experience. Organizations cannot simply purchase a software program, produce a set of posters filled with buzzwords, require their people to attend an hour of slide presentations, and instantly install an effective SMS. As with the development of any skill, it takes time, practice and repetition, the appropriate attitude, a cohesive approach, and constant coaching from involved mentors.

It is for this reason that a management framework that facilitates decision-making and shapes the environment in which employees' work is crucial to organizational performance in all aspects of the organization's business, including safety. A safety culture matures as safety management skills are learned and practiced and become second nature across the entire organization. The following have been found to be characteristics of organizations that consistently achieve safe results. These could be considered sub-culture behavior characteristics of a positive safety culture.

Open Reporting

Policies and processes can foster open reporting while, at the same time, stressing the need for continuous diligence and professionalism. The organization should encourage disclosure of error without fear of reprisal, yet it should also demand accountability on the part of employees and management alike. One method in which this is done is through Aviation Safety Action Programs (ASAP). The objective of the ASAP is to encourage air carrier and repair station employees to voluntarily report safety information that may be critical to identifying potential precursors to accidents. Under an ASAP, safety issues are resolved through corrective action rather than through punishment or discipline. The ASAP provides for the collection, analysis, and retention of the safety data that is obtained. ASAP safety data, much of which would otherwise be unobtainable, is used to develop corrective actions for identified safety concerns, and to educate the appropriate parties to prevent a reoccurrence of the same type of safety event. An ASAP is based on a safety partnership that will include the FAA and the certificate holder, and may include a third party, such as the employee's labor organization. To encourage an employee to voluntarily report safety issues, even though they may involve the employee's possible noncompliance with 14 CFR, enforcement-related incentives have been designed into the program. Whether an ASAP is developed or the organization develops their own confidential employee reporting system, they are essential components in assuring safety. They provide employee feedback for identifying new hazards and revising procedures.

Justness

The organization that is Just takes a proactive approach toward error disclosure yet demands accountability on the part of employees and management alike. The organization engages in identification of systemic errors through root cause analysis and implements preventative corrective action. It exhibits intolerance of undesirable behavior (recklessness and willful disregard for established procedures). An organization needs to communicate clearly to their employees that they will be treated fairly. It is a positive relationship in which the company is expected to communicate and effectively train their employees in regard to the policy and standard operation procedures, and the employees are expected to adhere to those policies and procedures follow them even though they feel there is a way that is more convenient or effective.

Openness

The organization that is open encourages and even rewards individuals for providing essential safety-related information that will improve the operation. This behavior can be feigned. Management will often encourage openness, such as the typical "open door policy," but once the safety concern has been received. Often the reply is why we don't do that, or the individual is belittled for bringing such a topic up. Management may accept the information, but then will do nothing to follow up with the concern or with the individual who brought it forth.

Another equally detrimental reaction by management is to inundate those with a concern with multiple barriers or additional tasks, making it nearly impossible for the process to be accomplished. Unfortunately, the result is an old email that never gets answered or resolved. You end up with management blaming labor, and labor blaming management for the inaction. Organizations who play these games have a way of looking open but not really being open. It may be certain departments, or a problem with the whole organization; and it usually stems from evident lack of support of SMS from leadership and or their effectiveness in communicating it with the managers.

When this happens a closed culture is created. A closed culture is one in which it appears to be open on the surface where the accountable executive has indicated his commitment, but there is an underlying attitude from the managers that reflects little support or buy-in for the use of the behaviors and processes encouraged under SMS. For example, the V.P. of aviation safety contacts a manager related to a series of concerns that are noted within the safety reports. The management, tired of one more email from Safety, either fail to respond or, if he does respond it is just a "Thank you for the email, yes we are aware of these issues and have taken steps to address those issues. Thanks again for your input." Three months later, nothing will have changed.

Use of Information

Effective use of all safety information assures an organization will have informed management decision-making. As your organization becomes more adept at collecting data, it may acquire an appetite for more and more data. That's normal. Information is valuable, but the resources and personnel must be provided so that it can be organized, categorized, analyzed, assessed and disseminated to those individuals who make the decisions. There are many off-the-shelf programs available, even as simple as a Microsoft Excel spread sheet shared with leadership. Regardless, information does not make the decision for leaders and managers, but helps them make a better decision.

This is an important distinction, because management can begin to feel that unless they have a huge data set in front of them, they can't make a decision. This is not the case. Experience gained from talking to pilots, mechanics, line personnel can be enough to make a decision, but the data may be starting point to determine what additional information the key safety management need in order to make a complete informed decision.

For example, as you look out on the ramp, you notice that most airplanes are taxing faster than you think is safe. Your initial reaction may be to send out an email, but instead you validate your concern with the flight data on taxi speeds on the ramp. The data indicates that 80% of the airplanes are taxing above the speed clearly outlined in company policies.

So you were right, they are going too fast. Although, as you look more closely at the data, you realize it will take more than an email to get them to slow down. In fact, many more questions need to be answered before you even make a decision regarding taxi speeds. Are 80% of the pilots willfully violating policies? Or do 80% of the pilots not know you have a policy? How did you train the pilots on taxi speeds? If they do know the policies, why aren't they following them? Is the speed too slow for the aircraft? Does management know of this issue, and if so, do they allow it or even encourage it? As you can see, data can help in the decision making by validating what you already know to be true; it can also help you ask the correct questions to get to the root cause of the unsafe acts that are manifesting as high taxi speeds, allowing you make the correct, fact-based decision.

Commitment to Risk Reduction

The organization expects direct management involvement in identifying hazards and managing risk. Expectations of management must be more than just what is published in a manual. Key safety management personnel, including the accountable executive, must expect direct management involvement. If the accountable executive, has directed a change to the system, he could ask that from management that a risk assessment be done and a brief on what strategies we will be used to mitigate the risk. This type of buy-in will raise the level of organization involvement in the actual reduction of risk and the processes to effectively achieve it.

Vigilance

This is the process that provides vigilance of ongoing operations and the environment to ensure the effectiveness of risk controls and an organizational awareness of emerging hazards. Remember, safety programs normally come to a halt after risk mitigation strategies have been determined, but what they often lack is vigilance. Vigilance is what happens after a risk mitigation strategy has been decided and before an incident or accident has occurred. For example, let's say that your organization experiences a gear up landing. Through the process, you determine that an observation program for all instructors needs to be instituted, as well as a formal training program for new hires.

Now what? As an organization are you vigilant? Do you follow up to see if the work was done? Can you see what the unintended consequences are? Can you see if the changes make a difference or not? Likewise, if you haven't encountered a gear up in one year, setting a new record are you aware that the time to be vigilant is now? What indicators should you be looking for in safety reports and through flight data monitoring that may indicate an eminent incident or accident? Who is vigilantly looking for this and analyzing the data? The organization must be committed to providing the resources and personnel necessary to be vigilant to reduce risk and to improve safety.

Flexibility

The organization uses information effectively to adjust and change in an effort to reduce risk. All aspects of the organization are under constant review and adjustment to meet changing demands. An organization must be willing to change to be successful. Doing it the way we've always done it, continues to be a slogan of leaders and managers who are lazy, fearful, and mediocre. We take risk with change, but it is measured, controlled, and calculated risk that has been thought through and planned out. Change involves decisions, and as stated earlier, the best decisions are made with the proper information. As we continually review our organization, there needs to be processes that help us identify the hazards across the organization to ensure we don't have a piece of the information but all the useful information we can. As previously discussed, this is where the safety performance meetings with the accountable executive and key safety management personnel from the different departments must come together. Realistically most significant hazards cross over multiple department lines and cannot be fully realized unless all are present to discuss.

Learning

The organization learns from its own failures and those of similar operations. What is important is the behavior characteristic of learning is more important than blaming. The fact that we are willing to learn says something about what we think about our employees as well. A learning organization realizes its employees will make mistakes; they won't always perform their jobs the right away. When mistakes are made, normally it is because of many factors that were outside of the control of the individual where the failure occurred. Employees recognize that what happened to this one individual could realistically happen to them as well. In fact, the one who made the mistake is probably now the most unlikely employee to make the mistake again.

The learning organization says, "help us so we don't make the same mistake you did." It turns someone who feels ashamed of making a mistake into someone who feels supported regardless of their mistake. It makes one who feels worthless into someone of worth who has information of value for his or her fellow employees.

Under SMS in a learning organization, when someone has a significant incident or accident, it is not unusual for them to volunteer to actually stand up in front of their peers and talk about what happen, how they contributed, and what could have been done differently, followed by a key safety management personnel following indicating changes made to the organization to help reduce the chances of the same thing happening again. A Learning organization requires a lot of trust between leadership and employees. It isn't something you can require, but something that is fostered through a consistent attitude of desiring to be better as an organization.

Training

Leadership actions, not just words, must show an investment in training, in up-to-date equipment, and tools so that employees are able to do their job safely. An investment in training is an investment in people. You are communicating, we want to you be competent, we want you to be comfortable with our policies, procedures, and processes so you can succeed. If you are committed to training, it is more than an annual initial training, but it is a recurrent planned activity. An investment must also be directed towards the trainers, ensuring you have qualified trainers and as an organization you are mentoring those who will be replacing your current trainers over time. One of the first places you will identify your organization as an entity that values training is when you start SMS. SMS requires that you have a training plan in which each person's training is commensurate with the position for both initial as well as recurrent rounds of training.

Accountability

To foster the development of a mature organization with a positive safety culture, an accountable executive must be in place. The accountable executive is the person who is the final authority over operations, controls, financial and human resources and retains ultimate responsibility for safety performance of the operation. While accountability starts at the top, SMS promotes action at all levels. In fact, if you see hazards or ways to reduce risk within your area of influence, there is an expectation for you to reduce that risk to the lowest practical level. SMS empowers all of the management staff, at all levels, which should convey, enhance and emphasize the organization's safety policy through exemplifying the policy in their daily work and in their one-on-one leadership styles. Decision-making should be kept at the lowest level appropriate to the complexity and criticality of the decision. Line managers are the people that own the process. They are in the best position to make appropriate changes. Senior management, including the accountable executive, should monitor actions and provide guidance.

Management Involvement

Management leadership should demonstrate their visible commitment to and involvement in safe operation performing their daily work. SMS processes do not have to be expensive or sophisticated; however, active personal involvement of operational leaders is essential. Those managers who "own" the processes in which risk resides must be the ones who accomplish safety management. Safety cultures also cannot be "created" or "implemented" by management decree, no matter how sincere their intentions. Every organization has a safety culture. It is embodied in the way the organization and its members approach safety in their jobs. If positive aspects of culture are to emerge, the organization's management must set up the policies and processes that create a working environment that fosters safe behavior. That is the purpose of the SMS processes.

Management involvement at all levels is demonstrated by:

- Formal risk analysis and resource allocation, as needed to assure mitigation of high consequence, high probability risks;
- Management action beyond rhetoric, actively involved in the decision making processes and participate in safety activities; and
- Strong safety assurance, combined with safety data analysis processes, yielding information, is used to drive risk reduction. An informed organization can take appropriate action to prevent accidents. Internal and external audits provide assurance that processes are working as designed and continuing to be effective. While it is possible to have a positive safety culture without a formal SMS, a strong safety culture can be fostered by the implementation of an effective SMS. The constant attention, commitment, and visible involvement provided by all levels of management, combined with continuing data analysis, SA activities and daily application of risk analysis and control techniques drive the organization toward safety culture maturity.

Informing

Related most closely with safety promotion component, informing is the behavior characteristic of informing or having an informed organization. As an organization gathers data and analyzes it, what do they do with it, and how it is communicated to the technical experts so that change can occur? An informed organization recognizes that those employees who are informed will be much more cognitive of safety and identifying hazards. Organizations must train as well explain changes that were made, why they were made and even how it will help to reduce risk. An employee who understands why, often has less trouble following the policy then one who doesn't understand. Information is key. An individual who is informed that there have been six runway incursions in the last month will behave differently than someone who was unaware of the hazards and higher risk items within the organization.

Remember, safety management is a learned skill. Organizations do not simply adopt a software program or a set of posters and buzzwords, attend an hour of slide presentations and instantly install an effective SMS. As with any skill, it takes time, practice, repetition, the appropriate attitudinal approach and good coaching. The safety culture matures as safety management skills are learned and practiced. The safety culture becomes second nature across the entire organization as trust builds and the organization functions as a team.

FAA Order 8000.369 identifies the following attributes of a positive culture of safety

- Competent personnel who understand hazards and associated safety risk are properly trained and have the skill and experience to ensure safe products/services are produced.
- Individual opinion is valued within the organization and personnel are encouraged to identify threats to safety and seek the changes necessary to overcome them.
- An environment where people are encouraged to develop and apply their skill and knowledge to enhance safety.
- Processes to analyze information from employees' reports, assess their content, develop actions as necessary and communicate results to the workforce and the public.
- Effective communications, including a non-punitive environment for reporting safety concerns.
- There are clear standards of behavior where there is a commonly understood difference between acceptable and unacceptable actions.
- Adequate resources to support the commitment to safety.
- A process for sharing safety information to develop and apply lessons learned with regard to hazard identification, safety risk analysis and assessment, safety risk controls, and other safety risk management responses. Sharing of information related to corrective actions, and results of management reviews are encouraged.
- Safety is a core value of the organization that endures over time, even in the face of significant personnel changes at any level.
- Willingness to recognize when basic assumptions should be challenged and changes are warranted—an adaptive and agile organization.
- Decisions are made based on knowing the risk involved in the consequences of the decision.

Safety culture is descriptive of organizations where each person involved in the organization's operations recognizes and acts on his or her individual responsibility for safety, and actively supports the organization's processes for managing safety. The outcome is that the organization's ability to manage safety continues to improve because decision makers at all levels work to use their knowledge of safety risk to learn and adapt, thus improving the system's ability to support safety outcomes.

Remember, having a safety program is not having a safety culture. These are two distinct things that are quite independent from one another. Organizations can have a great safety program that identify risk and other characteristics you would see with a positive safety culture, but still, they are not the same. When an organization finds out through a survey or audit that they do not have a positive culture in relation to safety, often they want to change it overnight.

You cannot create a positive culture in relation to safety overnight. It takes time. It takes between three to five years at a minimum, and from the accountable executive down, the change must be intentional and leadership must remain vigilant.

MEASURING YOUR POSITIVE SAFETY CULTURE

It is clear that a positive safety culture involves all levels of management as well as the technical experts, namely, the employees. Every organization has a culture; the question becomes is your culture positive and is it positive in regard to safety? One of the best ways to measure your safety culture is through a safety culture survey (see the Reader Resources for examples). Your first survey is free. It is just indicating where your culture is. Don't think of it as too low or just right, but instead identify where it is. At this point, your goal is to enhance your existing culture to be a more positive safety culture. Identify where you are at and begin to take small deliberate steps to enhance and improve your culture. Each year expand your survey, not only organizationally, but also departmentally. Track the change, the improvement, continue to learn, to be vigilant, willing to be flexible and change showing your commitment to reduce risk. As you continue this long process you will begin to change the culture, trust will replace fear. Open communication will replace reservations and silence.

SCENARIO DISCUSSION

Review each scenario and discuss how it does or doesn't relate to a positive safety culture.

1. An external audit was being conducted by a third party of a large cooperate flight department. During the audit, the chief pilot was asked, "How do people let you know if there are safety concerns within the organization?" The chief pilot proudly responded, "That's easy, I have an open door policy, if anyone has any safety concerns they can talk directly to me, anytime."

2. A group of frustrated pilots were given an hour on the schedule to meet with the accountable executive. All the pilots were uneasy and were afraid to tell the accountable executive their problems. The accountable executive could feel the uneasiness and he wanted them to feel comfortable. They began to explain that the standards office wasn't doing its job. The accountable executive responded, "You need to give me details if you expect me to do anything about it." The group seemed even more uncomfortable, but after some additional prodding from the accountable executive, a pilot said, "Actually there is an individual in standards whose flight checks are unfair." The accountable executive, feeling like he was getting somewhere, told the entire group of pilots, "Well tell me his name and we can take care of this problem quickly!" After that, the pilots quit giving input.

3. During an external audit, the SMS experts passed out one piece of paper to each of the employees in the room. They continued to instruct all the employees to write down where they thought the next incident or accident within the company would be. One employee asked if they had to write their name on the paper; another asked if their manager would see this; and others indicated that they just didn't feel comfortable writing anything.

REVIEW QUESTIONS

1. Define a positive safety culture.

2. What are three different ways an organization could establish a confidential employee reporting system that was indicative of a positive safety culture.

3. The winds are forecasted to be 35 knots up to the time of arrival, which is within the company's policies for wind limitations. En route, some un-forecast winds cause the winds to peak at 40 knots. The pilot determines the safest outcome would be to land in 40 knot winds, rather than risk attempting to land at another airport outside the area. How do you anticipate management will handle this situation under a positive safety culture?

4. Describe in your own words what a positive safety culture would look like in context of your organization.

5. What are the behavioral characteristics of a positive safety culture?

6. What do you think the FAA could not mandate that an organization of a positive safety culture?

Read the NTSB Accident Report NTSB/AAR-14/03 PB2014-108877 "Crash Following Encounter with Instrument Meteorological Conditions After Departure from Remote Landing Site Alaska Department of Public Safety Eurocopter AS350 B3, N911AA Talkeetna, Alaska March 30, 2013" and answer the following questions.

7. Identify the contributing factors of the accident.

8. Identify how safety culture played a role within the accident.

9. What behavior characteristics (listed in the text) of a Positive safety culture were not evident within the organization?

10. What attributes of a Positive safety culture (as listed in text) were not evident at this organization.

11. What future action will have to take place to begin to enhance the safety culture that existed at the time of the accident?

CHAPTER 14
Creating Your SMS Manual

OBJECTIVES

- To analyze the various components and resources to develop an SMS manual.
- To understand the requirements by 14 CFR Part 5 and SMSVP guide in regard to SMS records and documentation.
- To synthesize the information provided in various chapters to effectively develop an SMS manual.
- To develop the skill to evaluate an SMS manual and determine if it meets 14 CFR Part 5 and SMSVP guidance.

KEY TERMS

- Corporate Knowledge
- Elements
- Outputs

INTRODUCTION

While this SMS textbook in not an exhaustive repository of SMS principles and tools, it does provide the necessary information to establish and maintain a safety management system and a SMS manual. As you work through the process of transforming how your organization will manage safety, the development of your SMS manual can be used to chronicle your SMS journey, documenting your organization's path through the changes to full active conformance with SMS principles and FAA regulation/SMSVP guidelines. You will identify established policies, procedures, and processes you already have scattered through your organization in various manuals. You will also document corporate knowledge, and write the new policies, procedures, and processes that will guide your organization towards a more positive safety culture that is consistent with SMS principles.

REQUIREMENTS FOR DOCUMENTATION

For those organizations that have already made efforts to promote safety, whether through an established safety program or through a positive safety culture, the SMS requirements will identify how much within your organization is **corporate knowledge**. Corporate knowledge is the collective unwritten knowledge and experience that exist and are generally accepted by the stakeholders within an organization. When corporate knowledge is present, you will hear phases such as "we've always done it that way," "everyone knows that's how we do it." You might also hear, "Yes, it's written somewhere."

A simple method to identify corporate knowledge is to ask how a certain procedure or process is done, write it down, and then ask, "Where is this documented?" It can be disquieting to realize how much within your organization is undocumented or poorly communicated. As you practice this method, remember, seeing is believing. As you speak to different departments, managers will clearly understand what they expect; they will answer so clearly that it would be easy to assume the policy is written. Have them show you where it is written, even if they tell you it is written. This isn't an expression of lack of trust, but a process of validation. At times, policies, procedures, and processes are documented, but what is clearly written in the management's perspective is not clear to those who are either are unfamiliar with it or to those who are from the FAA. It is a lot less embarrassing to have these questions asked by your co-workers then by the FAA as you are being inspected for reaching SMS Active Conformance. Remember, what is documented must not only be understood by management, but, more importantly, it must be understood by an average employee who would be referencing or using the procedure or process.

Documentation

Having your policy, procedures, and processes documented is a cornerstone to establishing a safety management system. When someone does something contrary to policy, for example taxiing 20 mph on a congested ramp, think of the four components being ignored. Is there anything documented regarding the taxi speed for which everyone knows is too fast? How are the questions answered, such as:

- "Where is it written that we cannot taxi 20mph on the ramp?" *(Safety policy.)*
- "As an organization, how do we identify this is a reoccurring hazard?" *(Safety risk management.)*
- "If someone taxies fast or sees others not following the policy how can employees communicate this to management?" *(Safety assurance.)*
- "If they taxied too fast, did management communicate effectively to the employees regarding their expectations on taxi speeds and also provide training on those expectations?" *(Safety promotion.)*

While no one wants more documentation, it does play a vital role in a safety management system. Of course, documentation by no means corrects all safety concerns, in fact, too much documentation can have unintended negative consequences. Documentation, if done well, can, however, provide a clear framework for employee expectations and two-way communication that will assist in enhancing a positive safety culture.

Documentation is also clearly required by 14 CFR Part 5 and under the SMSVP. Job aids used by the FAA have two general areas, design and performance. These job aids are located in the SMSVP guide giving those seeking SMS active conformance the questions that will be asked by the FAA once your safety management system has been implemented. Design job aids are concerned with documentation and design of procedures and processes. Performance job aids are concerned with whether or not the documented and designed procedures and processes are actually performing the way they were designed. Therefore, performance design requires demonstration/validation of what has been designed. For example, Attachment 2 of the SMSVP is Subpart B Policy Design Validation Data Collection Tool (DCT) and Attachment 6 is "Safety Policy Demonstration DCT."

Under Attachment 2 a Yes or No question asks the following:

Minimum Requirements

Does the certificate holder's SMS have a safety policy that includes at least the following minimum requirements?

1. The certificate holder's safety objectives;
2. A commitment to fulfill the organization's safety objectives;
3. A clear statement to commit the necessary resources for implementation of the SMS;
4. A safety reporting policy that defines requirements for employee reporting of safety hazards or issues;
5. A policy that defines unacceptable behavior and conditions for disciplinary action; and
6. An emergency response plan that provides for the safe transition from normal to emergency operations in accordance with the requirements of §5.27, Coordination of Emergency Response Planning?

Under Attachment 6, the Yes and No questions are more in-depth questions that require validation, to the extent necessary, that the organization's Safety Policy has been conveyed throughout the organization. For example, the questions would be asked through interviews with employees:

1. Do employees at all levels of the organization demonstrate awareness of their system for employee reporting of safety hazards or issues?
2. Do employees at all levels of their organization demonstrate awareness of unacceptable safety behavior and conditions for disciplinary action?
3. Do employees at all levels of the organization demonstrate awareness of their defined safety accountability (i.e. can they relate safety objective(s) to their job)?

The FAA has established clear guidelines as to the requirement for documentation. 14 CFR Part 5 Subpart F – SMS Documentation and Recordkeeping identify the regulatory requirements and SMSVP Attachment 16 establishes the following additional detail that can help an organization develop their procedures and processes.

Records Retention

Certificate Management Teams (CMTs) will validate and continuously monitor, to the extent necessary, that the organization has record retention capability conforming to the standards (14 CFR Part 5 or SMSVP) in either paper or electronic media. This will include, but not be limited to, determining the ability of the organization to also retrieve archived records. For example, the CMT may ask the organization to show them a completed SRA on a past incident or accident; or if there was a hazard identified, the FAA may ask for a copy of the proactive SRA developed to address this hazard.

As your organization develops their SMS manual and begins document to meet the standard, the following questions must be considered:

- Have you applied its document design requirements to system operations? Remember your documentation must address the entire system, not just certain departments of the organization.
- Are the organization's personnel trained and have the competencies to perform their safety management related duties and responsibilities (qualification, training, knowledge, and experience);
- Are the process area/department personnel appropriately applying the documented process and procedures approved?
- Is what you documented actually what occurred or was applied?

The certificate holder is required to maintain records demonstrating conformance with applicable 14 CFR or SMSVP standards and provide historical reference documents for ongoing decision-making. The FAA will require the organization is capable of storing data for the required time periods defined in the standard and those required to retrieve stored data can do so in a timely manner. For paper records, access, protection from damage and misfiling are components of a good process. For electronic records, access, backup and protection from loss or overwrite are components of a good process. As per standard, know that the FAA will test the organization's record systems by requesting evidence that stored historical data matches the maximum retention period requirement. Remember the retention requirements established are the minimum; your organization could have more stringent requirements documented. If so, this is what the FAA will hold you accountable for.

For example: If today's date is 12/01/13 and there is a 24-month retention requirement, the certificate holder should be able to produce records from 12/01/11. If today's date is 12/01/18 and the retention requirement is unlimited, then records must be accessible back to the initial date of creation. If there is no "master record tracking document" defining the initial inception date of record, there is no standard to measure the historical completeness of a given record. It is difficult to determine if something is missing from recorded history if one does not know what is supposed to be in the historic file in the first place.

For example: for personnel records, the CMT should be able to pick individuals from actual surveillance activities and determine if they meet the required training outlined in the SMS manual. If there is a 24 month training records requirement, the FAA should be able to ask a management representative for employee records of individuals who have worked in the area for two years. If employees have SMS training modules in their job description, this should be documented in a training matrix detailing the requirement and process area.

Minimum Criteria outlined in 14 CFR §5.97 "SMS Records" are as follows:

> **(a)** The certificate holder must maintain records of outputs of safety risk management processes as described in subpart C of this Standard. Such records must be retained for as long as the control remains relevant to the operation.
>
> **(b)** The certificate holder must maintain records of outputs of safety assurance processes as described in subpart D of this Standard. Such records must be retained for a minimum of 5 years.
>
> **(c)** The certificate holder must maintain a record of all training provided a record of all training provided under §5.91 for each individual. Such records must be retained for as long as the individual is employed by the certificate holder.
>
> **(d)** The certificate holder must retain records of all communications provided under §5.93 for a minimum of 24 consecutive calendar months.

Minimum Criteria outlined in SMSVP Guide SMS Records are as follows:

- "Unlimited" record retention requirement. Records of SRM outputs for as long as the control remains relevant to the operation (each revision level of a process procedure should have SRM records from the date of original SMS acceptance).
- Five year record retention requirement. Safety assurance outputs (e.g., investigations, audits, corrective/preventive action, continuous process monitoring records (whether by day, week, or month) and employee hazard reports.
- Twenty-four month record retention requirement. Safety communications and safety management duties and responsibilities (hazard identification, hazard reporting, risk analysis, risk acceptance related to process area/department, etc.); and safety communications (e.g., the "why" documentation that includes bulletins, training records/curricula, corrective/preventive action records that require retraining of employees, meeting or briefing notes where "why" is explained, checklists of items reviewed at production meetings).

It is important to note, that having documents in an orderly fashion is commendable, but the records are not being used for their intended purposes, than records retention becomes a compliance drill instead of process that helps to management safety.

Maintaining records and record retention have many practical benefits to the management of safety.

1. Ensures all personnel are trained commensurate with their position.
2. Ensures that all SRA are completed and assure controls were effective.
3. System communication of the overall health of the organization.
4. Provide proper data and safety communication, including audit results, to the accountable executive in order to make fact-based decisions.
5. Historical records as to what was discussed in safety related meetings and briefings and "why" changes were made.
6. Historical records to assist in development of future SRA that are similar in nature.
7. Provide accountability in regard to "who" accepted a high or medium level risk item.
8. Demonstrate to the FAA you are meeting SMS records standards.
9. To help document and validate continuous improvement within the organization.

Regardless of the organizations size, it is important to identify the safety records manager(s) and perform enough validation activities to feel confident the organization is meeting the FAA records and documentation standards. Most importantly, an individual or department must be assigned to manage the SMS records. This management of SMS records would include:

1. Establishing and maintain procedures to identify, maintain, and dispose of SMS records;
2. Ensuring that all SMS records are legible, identifiable, and traceable to the activity involved; and
3. SMS records are maintained, readily retrievable, protected, and the retention schedule is adhered to.

As you develop your SMS manual it is also critical that you have procedures for ensuring all personnel are able to determine if the manual or form they are using is current. A common statement such as the one shown in Figure 14-1.

UNCONTROLLED COPY WHEN DOWNLOADED

Check the Master List to Verify that this is the Correct Revision Before Use

Figure 14-1 Example of a heading on an under-development SMS manual.

In additional to current documents, your organization should identify the individual(s) who are responsible for maintaining document(s) currency. This individual will most likely be a process manager or possibly a process owner who can identify and accept the risk for making a change to the procedure or process. Duties must be assigned regarding who will be responsible for ensuring outdated documents are promptly removed from all points of use or otherwise assured against unintended use.

Under the SMSVP Guide, there are several questions that summarize what the FAA will be asking at a minimum with the design phase. As with any design question, it should also be anticipated that you need to meet demonstration/performance expectation to determine if your organization has applied the processes documented.

- Does the certificate holder have a process to develop and maintain SMS documentation that describes their safety assurance processes and procedures?
- Does the certificate holder's SMS contain a process to maintain records of their safety assurance process outputs for a minimum of 5 years?
- Does the certificate holder's SMS require the organization have a process to maintain records of their SRM outputs for as long as the control(s) remain relevant to their operation, to include:
  ▷ Records of identified hazards or no hazard risk acceptance;
  ▷ Records of associated risks with identified hazards, as applicable;
  ▷ Records of analysis for each risk, as applicable; and
  ▷ Records of new risk controls approved to mitigate unacceptable risks, as applicable?

PLANNING FOR SUCCESS

On occasion, the authors of this book have received calls from a work-study group or a single individual from a smaller organization asking for help. Generally, it is someone who is relatively lower on the organizational chain that has been told to make a manual or get a SMS program going. While airlines, with safety departments, would not take this approach, smaller scale operations such as a flight school, air taxi, or airport often do, hoping for a quick solution. If you a considering such an approach, stop. It would be better to do nothing. Even though you may have individuals within your organization who have read the regulation, and even though you write a nice looking manual that meets all the "design" characteristics, when it comes time for demonstration/performance of your new system your entire uninvolved organization will be lost.

The Big Picture

SMS affects the entire system. If you expect the system to work, you will need to involve all parts. Consider leadership involvement from the different columns you established on your gap tool. For example, if your organization decided on four major columns, such flight operations, academics, maintenance, and safety, then invite individuals from each of these columns.

For example, a large Part 141 pilot training operation or an airport may begin by developing a safety council. A Safety Council group of leaders is comprised of the different departments, such as flight operations, maintenance, maintenance training, ground operations, line operations, training, and safety department, as well as the accountable executive. If you have a good relationship with the FAA, it would be our suggestion to even invite the FAA to participate as you form up this group. Frankly, the more you involve the FAA the better. Remember, the local FAA or CMT will be the individuals who need to validate compliance with your organization. The more they are involved with the development of your SMS, the easier it will be for them to understand the complexity of your organization and be able to validate compliance.

As you begin the development of your SMS, at a minimum, make an appointment with the FAA to gauge their desired involvement and willingness to participate. Under Part 5, the FAA is mandated to participate, while under SMSVP the FAA has not been provided additional funding. The safety council can start working together to create safety policy and promote safety throughout the organization. The initial development of a safety survey may be created to begin to measure the company's safety culture. As your safety council matures, sub-committees can be established, such as an SMS committee to begin to formulate your SMS manual.

This process will not be easy. Remember that the path to active conformance is a journey, even a marathon. The efforts to involve the entire organization in the early stage will pay dividends in the later stages. Ensure that your safety subcommittee regularly briefs the safety council so they are aware of the progress, the barriers, and resources needed to continue to move forward.

CREATING YOUR SMS MANUAL

Why create a SMS manual? FAA Part 5.95 SMS Documentation requires, "The certificate holder must develop and maintain SMS documentation that describes the certificate holder's":

1. Safety policy.
2. SMS processes and procedures.

It seems like a simple requirement that can be done within a few pages depending on the size of your operation. Again, the process is scalable. The University of North Dakota (UND) Aviation program's first SMS manual was just 40 pages long. It was simple and to the point. Instead of including everything within the SMS manual, UND referenced other existing manuals. For example, under UND's safety program, there was an emergency response plan already developed; therefore, when they developed the SMS manual they stated, "…As documented in the UND Aerospace Emergency Response Manual…" The regulations are vague to allow you as an organization to tailor them in various ways to meet the unique challenges of your organization, while still meeting SMS principles.

As different key aspects are discussed, remember reference your SMS Part 5 requirements, SMS AC120-92B, and the SMSVP Guide.

Safety Policy Statement

As previously discussed, the policy statement is a foundation of the SMS. You must first persuade your accountable executive to buy-in, and commit, in writing, to establish and maintain an SMS and provide the resources necessary to develop a SMS. This is critical to move forward on the SMS journey. It will be the first page in your SMS manual.

Developing the safety policy statement will not be easy. The most difficult part of the safety policy statement will be the safety objectives. The rest of the safety policy statement is verbiage committing key points that lay the foundation for a safety management system. The safety objectives are more complex and must be written and followed by the key leadership or process owners, including the accountable executive. These safety objectives will drive the SMS to measure progress and success. As the accountable executive has regular safety performance meetings, these safety objectives will be reviewed, updated, and evaluated. See Chapter 9 for additional information on safety policy beginning on page 120.

A corresponding task is to determine the key leadership personnel. These are the individuals who will meet 14 CFR §5.23 and §5.25, which state the following:

§5.23 Safety accountability and authority.

(a) The certificate holder must define accountability for safety within the organization's safety policy for the following individuals:

(1) accountable executive, as described in §5.25.

(2) All members of management in regard to developing, implementing, and maintaining SMS processes within their area of responsibility, including, but not limited to:

(i) Hazard identification and safety risk assessment.

(ii) Assuring the effectiveness of safety risk controls.

(iii) Promoting safety as required in subpart E of this part.

(iv) Advising the accountable executive on the performance of the SMS and on any need for improvement.

(3) Employees relative to the certificate holder's safety performance.

(b) The certificate holder must identify the levels of management with the authority to make decisions regarding safety risk acceptance.

§5.25 Designation and responsibilities of required safety management personnel.

(a) *Designation of the accountable executive.* The certificate holder must identify an accountable executive who, irrespective of other functions, satisfies the following:

(1) Is the final authority over operations authorized to be conducted under the certificate holder's certificate(s).

(2) Controls the financial resources required for the operations to be conducted under the certificate holder's certificate(s).

(3) Controls the human resources required for the operations authorized to be conducted under the certificate holder's certificate(s).

(4) Retains ultimate responsibility for the safety performance of the operations conducted under the certificate holder's certificate.

(b) *Responsibilities of the accountable executive.* The accountable executive must accomplish the following:

(1) Ensure that the SMS is properly implemented and performing in all areas of the certificate holder's organization.

(2) Develop and sign the safety policy of the certificate holder.

(3) Communicate the safety policy throughout the certificate holder's organization.

(4) Regularly review the certificate holder's safety policy to ensure it remains relevant and appropriate to the certificate holder.

(5) Regularly review the safety performance of the certificate holder's organization and direct actions necessary to address substandard safety performance in accordance with §5.75.

(c) *Designation of management personnel.* The accountable executive must designate sufficient management personnel who, on behalf of the accountable executive are responsible for the following:

(1) Coordinate implementation, maintenance, and integration of the SMS throughout the certificate holder's organization.

(Continued)

> (2) Facilitate hazard identification and safety risk analysis.
>
> (3) Monitor the effectiveness of safety risk controls.
>
> (4) Ensure safety promotion throughout the certificate holder's organization as required in subpart E of this Standard.
>
> (5) Regularly report to the accountable executive on the performance of the SMS and on any need for improvement.

Reading these regulations can be intimidating. The important thing is there are sufficient personnel to carry out the essential requirements of SMS. Don't think of each subheading representing a different role from a different individual, but instead look at them as an overview of the primary job duties of one or more individuals. A large organization, perhaps having 100 aircraft, , may require multiple individuals, pulling resources from different departments, while a small airport may combine all of these tasks into one person who answers directly to the accountable executive.

Regardless of size, SMS can be summarized as the accountable executive indicating to key leadership that their job performance will be based on more than just production but also on safety objectives that they must update and strive to meet and exceed.

Before trying to decide every job function, finish your safety policy statement and then be begin working on your gap analysis tool and implementation plan.

Gap Analysis Tool and Implementation Plan

On the Reader Resources page for this book, you can download and use a FAA gap analysis tool. As you recognize these gaps, you will want to first develop an implementation summary plan. The summary does not need to have all the details of whom, how, what and when, but the Implementation Summary will answer at least three questions for each gap you identify:

1. You will be responsible for filling in the gap?
2. What is the title of the person responsible?
3. What date will it be complete by?

As you begin working through the SMS gap analysis tool, note that the questions will look very similar under each of the four components. While these questions seem similar in verbiage, the context changes dramatically between each component and how you answer the question.

For example, the gap tool may be asking about safety reporting under each component. While the question is similar under Safety Policy, the FAA wants to know if it is a written process. Under Safety Risk Management the FAA is concerned with how it is used to analyze risk. Under Safety Assurance, the FAA is concerned with how it is used to assure risk mitigation strategies are implemented and working as intended; and finally, under Safety Promotion, the FAA is concerned with whether people are aware of how to conduct safety reporting and how that information is shared with the entire organization. As you create your processes, you must always be considering how they are impacted by all four components.

Once your SMS committee or working groups from different departments complete the gap tool and the Summary implementation plan you can use the information to begin to create sections of your SMS manual. For those areas that are documented elsewhere you may elect to move them to your new SMS manual, or you may choose to just reference them in your SMS manual. Start documenting areas that are corporate knowledge in your manual. These will be easy wins, because they are generally accepted throughout the organization. For example, everyone keeps current forms in a certain network drive. Document where you keep those forms and who is responsible for them. Where your gaps are the largest you will have to systematically begin the process of filling in the gaps and involving the personnel most intimate with the each department.

Have one person, deemed the SMS planner, or small groups orchestrate the various planning meetings that will need to occur to produce organizational buy-in. The SMS planner will also be the individual who maintains communication with the FAA.

The end goal of your implementation plan is to bring you to active conformance. At that point all processes of SMS are documented and designed within your organization, and implemented, too.

Written policy will be in place that the accountable executive expects to be used. Implementation does not mean your processes have yet been internally or externally audited to ensure they are operating as designed. This is an important distinction: active conformance can occur with relatively new procedures, but once Active Conformance has been obtained your local CMT will be conducting continuous oversight to assure the processes are being used by the organization.

FAA INVOLVEMENT

This can be approached in two ways:

1. You can develop your manual, use the FAA job aids to verify performance of your SMS then pursue SMS recognition; or
2. Involved the FAA from the very beginning and systematically work through the various levels (active applicant, active participant, active conformance)

Part 121 operators must involve the FAA sooner than later. Under Part 5, Part 121 operators are on a time line to have an implementation plan developed as well as regulatory compliance obtained. Other certificate holders have more freedom. The benefit of early involvement from your local FAA is buy-in and they see your work and progress. Also for motivation in the long SMS marathon, you can celebrate the small successes of reaching active applicant and well as active participant.

Most importantly, make your manual user friendly to the FAA and your developers. Adopt the FAA numbering system, which allows you to identify the SMS manual section or process that corresponds with the 14 CFR, SMSVP, and the gap analysis tool. If your SMS Safety and 14 CFR Policy are both §5.21 compliant, you make it simple for the FAA, internal evaluators, and external auditors to identify your SMS design.

EVALUATING SUCCESS

Having ways to measure your success is critical in SMS. If you don't know where you are at, or where you have been, it is logically hard to determine if you have improved or are even successful at SMS. Use your Safety objectives to begin to determine what you want to measure and how.

Once you've completed your SMS Manual design and you are using the document you will feel successful, but in one sense this is where the real work begins. Everyone will need to be trained in the ways of the manual commensurate with his or her position. From the basic employee to the CEO, you will need to identify what they need for initial training as well as what and how often they will need recurrent training. You will experience pockets of success as well as pockets of regression within your organization as you struggle to implement the SMS Manual's dictates. Remember, at first, just measure it. The old cliché is appropriate, "where ever you go, there you are," meaning: you have to start somewhere. Aim for improvement, always moving toward a safer, better organization.

REVIEW QUESTIONS

1. Give an example of corporate knowledge at your work place.

2. List three practical benefits of records management and retention.

3. What is the different between a procedure and a process?

4. How could the SMSVP job aids be helpful to a Part 121 carrier or other certificate holder?

5. How long must training records be saved?

6. Give two examples of safety communication that must be kept and how you would keep them in your organization.

7. What are four responsibilities of an SMS accountable executive?

8. What is one positive aspect of having multiple areas or departments identified when using the gap tool?

9. What is a negative impact of having a large number of areas identified for using the gap tool compared with having a smaller number?

10. What are the three minimum requirements for a summary implementation plan?

GLOSSARY

5-STEP PROCESS. Defined within SRM: (1) define the system, (2) identify the hazards, (3) assess the risks, (4) analyze the risks, and (5) mitigate the risks. In this guidebook, safety risk assessment is used exclusively when describing the 5-step process and the conduct of that process with a panel of SMEs.

FAA AIR TRAFFIC ORGANIZATION (ATO). The operational arm of the FAA. It is responsible for providing safe and efficient air navigation services to 30.2 million square miles of airspace. *(http://www.faa.gov/about/office_org/headquarters_offices/ato/)*

FAA OFFICE OF AVIATION SAFETY (AVS). Responsible for the certification, production approval, and continued airworthiness of aircraft; and certification of pilots, mechanics, and others in safety-related positions. *(http://www.faa.gov/about/office_org/headquarters_offices/avs/)*

ACCEPTABLE RISK. The level of risk that individuals or groups are willing to accept given the benefits gained. Each organization will have its own acceptable risk level, which is derived from its legal and regulatory compliance responsibilities, its threat profile, and its business/organizational drivers and impacts.

ACCOUNTABLE EXECUTIVE. The single identifiable person who has the ultimate responsibility for the effective and efficient performance of the organization's safety management system.

ACTIVE FAILURE. An error of omission or commission that is made in the course of a particular operation. An active failure can also be a known problem or a known mechanical deficiency or fault.

AIRCRAFT ACCIDENT. An occurrence, associated with the operation of an aircraft, that occurs between the time any person boards the aircraft with the intention of flight and until all such persons have disembarked; and in which any person suffers death or serious injury, or the aircraft receives substantial damage.

AIRCRAFT INCIDENT. An occurrence, other than an accident, that is associated with the operation of an aircraft and that affects or could affect the safety of operations. An aircraft incident is a near miss episode with minor consequences that could have resulted in greater loss. An unplanned event that could have resulted in an accident, or did result in minor damage, and indicates the existence of, though may not define, a hazard or hazardous condition.

AS LOW AS REASONABLY PRACTICABLE (ALARP). Describes a safety risk reduced to a level that is as low as reasonably practicable; that is, any further risk reduction is either impracticable or grossly outweighed by the cost. (ICAO Safety Management Manual) Note: in the latest version of ICAO Safety Management Manual (3rd Edition, 2013) ALARP was removed. It is provided here for reference only.

AUDIT (SMS AUDIT). A review of an organization's SMS program to verify completion of tasks and determine an organization's compliance with FAA directives and procedures.

COMMON CAUSE FAILURE. A failure that occurs when a single fault results in the corresponding failure of multiple system components or functions.

COMPARATIVE SAFETY ASSESSMENT (CSA). A safety analysis that provides a list of hazards associated with a project proposal, along with a risk assessment of each alternative-hazard combination. A CSA is used to compare alternatives from a risk perspective.

CONFORMANCE. Objective evidence showing the agreement in nature or form of a presented document, process, or system.

CONSTRUCTION SAFETY AND PHASING PLAN (CSPP). A document that outlines procedures, coordination, and control of safety issues during construction activity on an airport.

CORPORATE SRM. A process to identify hazards and associated risks, analyze risks, and develop new risk controls affecting multiple process owner areas/departments within the organization. Final risk acceptance for Corporate SRM may be accomplished at a management level above the process owner/department level or by a committee.

CREDIBLE. It is reasonable to expect that the assumed combination of conditions that define the system state will occur within the operational lifetime of the component/event.

DEPARTMENTAL SRM. The process used to identify hazards and associated risks, analyze risk, and develop new risk controls affecting a single process owner area/department within the organization. Final Risk acceptance for departmental SRM may be accomplished by the appropriate process owner or manager.

DESIGN REVIEW. An auditing or evaluation process which determines if a certificate holder's safety management processes conform to the Safety Management System Voluntary Program (SMSVP) Standard.

ERROR-TOLERANT SYSTEM. A system that is designed and implemented in such a way that, to the maximum extent possible, errors and equipment failures do not result in an incident or accident. An error-tolerant design is the human equivalent of a fault-tolerant design.

EXISTING CONTROL. A mitigation already in place that prevents or reduces the hazard's likelihood or mitigates its effects. A control can only be considered existing if it has been validated and verified with objective evidence.

FAA OFFICE OF AIRPORTS (ARP). Provides leadership in planning and developing a safe and efficient national airport system. The office is responsible for all programs related to airport safety and inspections and standards for airport design, construction, and operation (including international harmonization of airport standards). The office also is responsible for national airport planning and environmental and social requirements and establishes policies related to airport rates and charges, compliance with grant assurances, and airport privatization. (*http://www.faa.gov/about/office_org/headquarters_offices/arp/*)

FAIL OPERATIONAL. A system designed such that if it sustains a fault, it still provides a subset of its specified behavior.

FAIL SAFE. A system designed such that if it fails, it fails in a way that will cause no harm to other devices or present a danger to personnel.

FATAL INJURY. Fatal injuries include all deaths determined to be a direct result of injuries sustained in the accident, and within 30 days of the date of the accident.

FIVE "M" MODEL. A model often used to help define an operational system, composed of five elements: Mission, Man (or huMan), Machine, Management, and Media (or environMent).

GAP ANALYSIS. The process where the organization compares existing processes and procedures, compared to the desired safety components and activities under the SMS or SMSVP program requirements. Gap analysis provides an organization an initial SMS development plan and roadmap for compliance.

HAZARD. A condition that can lead to injury, illness or death to people; damage to or loss of a system, equipment, or property; or damage to the environment. A condition that could foreseeably cause or contribute to an accident.

HAZARD ASSESSMENT. A systematic, comprehensive evaluation of a change, operation, system, or safety issue.

HIGH-RISK HAZARD. A hazard with an unacceptable level of safety risk; the change cannot be implemented unless the hazard's associated risk is mitigated and reduced to medium or low.

INCIDENT. An occurrence, other than an accident, associated with the operation of an aircraft that affects or could affect the safety of operations.

INTERNATIONAL CIVIL AVIATION ORGANIZATION (ICAO). A specialized agency of the United Nations, the ICAO promotes the safe and orderly development of international civil aviation throughout the world. (*http://www.icao.int/about-icao/Pages/default.aspx*)

KEY PERFORMANCE INDICATOR (KPI). A set of quantifiable measures that a company or industry uses to gauge or compare performance in terms of meeting strategic and operational goals. Within the context of SRM, the KPI will be safety related.

LATENT FAILURE. An error or failure whose adverse consequences may lie dormant within a system for a long time, becoming evident when combined with other factors.

LIKELIHOOD. The estimated probability or frequency, in quantitative or qualitative terms, of a hazard's effect.

MATERIAL CHANGE. Any change, relating to a construction project, that is a result of the environmental or design process and/or alternative selection that changes the physical layout. Such changes could introduce safety risks.

MINOR INJURY. Any injury that is neither fatal nor serious.

NATIONAL AIRSPACE SYSTEM (NAS). The common network of U.S. airspace; air navigation facilities; equipment and services; airports or landing areas; aeronautical charts and information services; rules, regulations, and procedures; technical information; and labor and material. The NAS includes system components shared with the military.

NONCONFORMITY. Non-fulfillment of a requirement. This includes but is not limited to non-compliance with Federal regulations. It also includes an organization's requirements, policies, and procedures as well as requirements of safety risk controls developed by the organization.

OBJECTIVE EVIDENCE. Documented proof; the evidence must not be circumstantial and must be obtained through observation, measurement, test, or other means.

OPERATIONAL RISK MANAGEMENT (ORM). A decision-making tool used by personnel at all levels to increase effectiveness by identifying, assessing, and managing risks. By reducing the potential for loss, the probability of a successful mission increases.

OUTCOME. An outcome is the potential undesirable result of a hazard or the ill effects potentially resulting from exposure to a hazard. A specific system state and sequence of events supported by data and expert opinion that clearly describes the outcome. The term implies that it is reasonable to expect the assumed combination of conditions may occur within the operational lifetime of the system. Note: Other terms used in risk management as substitutes for outcome included consequence, effect, and result. In this book, outcome is used rather than consequence or effect.

PERFORMANCE DEMONSTRATION. Determines how well a certificate holder's safety management activities work when applied to actual operational conditions.

PRELIMINARY HAZARD ASSESSMENT (PHA). An overview of the hazards associated with an operation or project proposal consisting of an initial risk assessment and development of safety related requirements. This can result in a Preliminary Hazard List (PHL); a list of anything that the analyst can think of that can go wrong, based on the concept, its operation, and implementation.

PROCESS. A series of steps or activities that are accomplished in a consistent manner to ensure a desired result is attained on an ongoing basis.

PROCESS MANAGERS. Individuals assigned by a process owner who can accept the appropriate level of risk within their area of responsibility. Process managers' duties shall include: hazard identification, safety risk assessment, and risk acceptance; evaluating the effectiveness of safety risk controls; promoting safety; and submitting performance reports to the Key Safety Management/Process Owner(s) on SMS performance.

PROCESS OWNERS. The key safety management personnel who oversee specific areas of operations.

QUALITATIVE DATA. Subjective data that is expressed as a measure of quality; nominal data.

QUALITATIVE RISK. The level of risk based on subjective measures, rather than quantitative metrics.

QUALITY ASSURANCE. A program for the systematic monitoring and evaluation of the various aspects of a project, service, or facility to ensure that standards of quality are being met. It is a process to assess and review the processes and systems that are used to provide outputs (whether services or products) and to identify risks and trends that can be used to improve these systems and processes.

QUALITY CONTROL. A process that assesses the output (whether a product or service) of a particular process or function and identifies any deficiencies or problems that need to be addressed.

QUANTITATIVE DATA. Objective data expressed as a quantity, number, or amount, allowing for a more rational analysis and substantiation of findings.

QUANTITATIVE RISK. The level of risk based on objective data and metrics.

REDUNDANCY. A design attribute in a system that ensures duplication or repetition of elements to provide alternative functional channels in case of failure. Redundancy allows the service to be provided by more than one path to maximize the availability of the service.

RESIDUAL RISK. (1) The remaining predicted severity and likelihood that exist after all selected risk control techniques have been implemented. (2) The level of risk that has been verified by completing a thorough monitoring plan with achieved measurable safety performance target(s). Residual risk is the assessed severity of a hazard's effects and the frequency of the effect's occurrence.

RISK. The composite of predicted severity and likelihood of the potential effect of a hazard. Hazards present risk. Risk is the composite of predicted severity and likelihood of the potential outcome of a hazard. Risks may be categorized as follows:

- Initial—the severity and likelihood of a hazard's risk when it is first identified and assessed, including the effects of pre-existing risk controls in the current system.
- Current—the predicted severity and likelihood of a hazard's risk at the current time.
- Residual—the risk that remain after all risk mitigations have been implemented or exhausted and all risk mitigations have been verified.

RISK ACCEPTANCE. The confirmation by the appropriate management official that he or she understands the safety risk associated with the change and that he or she accepts that safety risk into the organization. Risk acceptance requires that signatures have been obtained for the safety requirements identified in the SRMD and that a comprehensive monitoring plan has been developed and will be followed to verify the predicted residual risk.

RISK ANALYSIS. The process during which a hazard is characterized for its likelihood and the severity of its effect or harm. Risk analysis can be either quantitative or qualitative; however, the inability to quantify or the lack of historical data on a particular hazard does not preclude the need for analysis.

RISK ASSESSMENT. The process by which the results of risk analysis are used to make decisions (Fact-based decision making). The process combines the effects of risk elements discovered in risk analysis and compares them against acceptability criteria. A risk assessment can include consolidating risks into risk sets that can be jointly mitigated, combined, and then used in making decisions.

RISK ASSESSMENT CODE (RAC). The ranking of risks based on the combination of likelihood and consequence (severity) values. A widely used SRM term throughout DoD and governmental agencies.

RISK CONTROL. (1) An action used to reduce or eliminate the risk severity and/or likelihood, via the application of engineering and/or administrative hazard controls. Risk control can also be anything that mitigates or lessens the risk. Note: The term risk mitigation is often used instead of risk control. See also existing control. (2) Safety risk controls necessary to mitigate an unacceptable risk should be mandatory, measurable, and monitored for effectiveness.

RISK MATRIX. Table depicting the various levels of severity and likelihood as they relate to the levels of risk (e.g., low, medium, or high). Risk matrices may be color coded; unacceptable (red), acceptable (green), and acceptable with mitigation (yellow).

1. Unacceptable (Red). Where combinations of severity and likelihood cause risk to fall into the red area, the risk would be assessed as unacceptable and further work would be required to design an intervention to eliminate that associated hazard or to control the factors that lead to higher risk likelihood or severity.

2. Acceptable (Green). Where the assessed risk falls into the green area, it may be accepted without further action. The objective in risk management should always be to reduce risk to as low as practicable regardless of whether or not the assessment shows that it can be accepted as is. This is a fundamental principle of continuous improvement.

3. Acceptable with Mitigation (Yellow). Where the risk assessment falls into the yellow area, the risk may be accepted under defined conditions of mitigation. An example of this situation would be an assessment of the impact of a non-operational aircraft component for inclusion on a minimum equipment list (MEL). Defining an Operational (O) or Maintenance (M) procedure in the MEL would constitute a mitigating action that could make an otherwise unacceptable risk acceptable, as long as the defined procedure was implemented. These situations may also require continued special emphasis in the Safety Assurance function.

RISK MITIGATION. Any action taken to reduce the risk of a hazard's effect.

ROOT CAUSE ANALYSIS. Analysis of deficiencies to determine their underlying root cause.

SAFETY. (1) Freedom from unacceptable risk *(FAA definition)*. (2) The state in which the risk of harm to persons or property damage is reduced to, and maintained at or below, an acceptable level throughout a continuing process of hazard identification and risk management *(ICAO definition)*.

SAFETY ASSURANCE. The process and procedures of management functions that evaluate the continued effectiveness of implemented risk mitigation strategies, support the identification of new hazards, and function to systematically provide confidence that an organization meets or exceeds its safety objectives through continuous improvement. Data-based decision are made through the collection, analysis, and assessment of information.

SAFETY EVALUATION. Procedures to monitor performance with respect to safety objectives, SMS requirements, and/or safety initiatives.

SAFETY MANAGEMENT SYSTEM (SMS). The formal, top-down, organization-wide approach to managing safety risk and ensuring the effectiveness of safety risk controls. It includes systematic procedures, practices, and policies for the management of safety risk.

SAFETY MANAGEMENT SYSTEM VOLUNTARY PROGRAM (SMSVP). The formal voluntary application of safety management system requirements.

SAFETY MARGIN. The buffer between the actual minimum-level requirement and the limit of the hardware or software system.

SAFETY OBJECTIVE. A measurable goal or desirable outcome related to safety.

SAFETY PERFORMANCE. Realized or actual safety accomplishment relative to the organization's safety objectives.

SAFETY PERFORMANCE INDICATOR (SPI). A data-based safety parameter used for monitoring and assessing safety performance.

SAFETY PERFORMANCE TARGETS. Measurable goals used to verify the predicted residual risk of a hazard. They should quantifiably define the predicted residual risk.

SAFETY POLICY. The organization's documented commitment to safety, which defines its safety objectives and the accountabilities and responsibilities of its employees in regards to safety. Safety policy defines the fundamental approach to managing safety that is to be adopted within an organization. Safety policy further defines the organization's commitment to safety and overall safety vision.

SAFETY PROMOTION. The combination of safety culture, training and communication of safety information to support the implementation and operation of an SMS in an organization.

SAFETY RISK. The composite of predicted severity and likelihood of the potential effect of a hazard.
- Initial—the predicted severity and likelihood of a hazard's effects or outcomes when it is first identified and assessed; includes the effects of pre-existing risk controls in the current environment.
- Current—the predicted severity and likelihood at the current time.
- Residual—the remaining predicted severity and likelihood that exists after all selected risk control techniques have been implemented.

SAFETY RISK ASSESSMENT. Assessment of a system or component, often by a panel of system subject matter experts (SMEs) and stakeholders, to compare an achieved risk level with the tolerable risk level.

SAFETY RISK MANAGEMENT. The process within the SMS composed of describing the system, identifying the hazards, and analyzing, assessing and controlling risk. A standard set of processes to identify and document hazards, analyze and assess potential risks, and develop appropriate mitigation strategies.

SAFETY RISK MANAGEMENT DOCUMENT (SRMD). Thoroughly describes the safety analysis for a proposed NAS change. The SRMD is an ATO-specified description of the safety analysis for a given proposed change. An SRMD documents the evidence to support whether or not the proposed change to the system is acceptable from a safety risk perspective. SRMDs are maintained by the organization responsible for the change for the lifecycle of the system or change.

SAFETY RISK MANAGEMENT PANEL. A group formed to formalize a proactive approach to system safety and a methodology that ensures hazards are identified and unacceptable risk is mitigated before the change is made. An SRM Panel provides a framework to ensure that, once a change is made, the change will be tracked throughout its lifecycle.

SERIOUS INJURY. An injury that (1) requires hospitalization for more than 48 hours, commencing within 7 days from the date the injury was received; (2) results in a fracture of any bone (except simple fractures of fingers, toes, or nose); (3) causes severe hemorrhages, nerve, muscle, or tendon damage; (4) involves any internal organ; or (5) involves second- or third-degree burns, or any burns affecting more than 5 percent of the body surface.

SEVERITY. The measure of how severe the results of a hazardous condition's outcome are predicted to be. Severity is one component of risk. The safety risk of a hazard is assessed on the combination of the severity of and the likelihood (probability) of the potential outcome(s) of the hazard.

SINGLE POINT FAILURE. A failure of an item that would result in the failure of the system and is not mitigated by redundancy or an alternative operational procedure.

SMS MANUAL. A document developed by an organization that describes the SMS components and how they will be established and will function. The SMS Manual describes how the organization's SMS functions.

SOURCE (OF A HAZARD). Any real or potential origin of system failure, including equipment, operating environment, human factors, human-machine interface, procedures, and external services.

STAKEHOLDER. A group or individual that is affected by or is in some way accountable for the outcome of an undertaking; an interested party having a right, share, or claim in a product or service, or in its success in possessing qualities that meet that party's needs and/or expectations.

SYSTEM. An integrated set of constituent pieces combined in an operational or support environment to meet a defined objective. Elements include people, hardware, software, firmware, information, procedures, facilities, services, and other support facets. See the also the Five "M" Model.

SYSTEM STATE. An expression of the various conditions, characterized by quantities or qualities, in which a system can exist.

TRIGGERS (FOR SRM). The requirements, precursors, or organizational plans that lead to initiation of the SRM process.

UNACCEPTABLE LEVEL OF SAFETY RISK. A high-risk hazard or a combination of medium/low risks that collectively increase risk to a high level.

VALIDATION. The process of proving the functions, procedures, controls, and safety standards are correct and the right system is being built (that is, the requirements are unambiguous, correct, complete, and verifiable.)

VALIDATION PLAN. The document which forecasts the resources needed to perform applicable confirmation of a certificate holder's safety management activities and processes.

WORST CREDIBLE EFFECT. The most unfavorable, yet believable and possible, condition given the system state.

INDEX

1:600 Rule *81*

14 CFR Part 5 *7, 11, 20–21, 44, 58, 68, 92, 106, 165, 170*
    §5.5 System analysis and hazard identification *28*
    §5.21 Safety policy *16–17, 191*
    §5.23 Safety accountability and authority *17, 189*
    §5.25 Designation and responsibilities of required safety management personnel *17, 189–190*
    §5.27 Coordination of emergency response planning *18*
    §5.51 Applicability *28*
    §5.55 Safety risk assessment and control *29*
    §5.71 Safety performance monitoring and measurement *34*
    §5.73 Safety performance assessment *34*
    §5.75 Continuous improvement *35*
    §5.91 Competencies and training *37*
    §5.93 Safety communication *37*
    §5.97 SMS records *185*

14 CFR Part 91 *50*

14 CFR Part 121 *58, 60, 65, 92, 106–107*

14 CFR Part 135 *58*

A
acceptable risk *25*

accountability *16, 123*

accountable executive *15, 121–123, 189*
    scalability *62*

active failure *72*

Air Traffic Service (ATS) *10*

audit *31, 160–167*
    auditing tools *160–163*

authority *15*

Aviation Safety Action Program (ASAP) *30, 32, 60, 155, 171*

Aviation Safety Reporting System (ASRS) *155–156*

aviation service provider *59*

C
certificate holder *94, 97*

Certificate Management Team (CMT) *8, 93, 96–97, 104*
    CMT VPP *101, 104*

civil aviation authorities (CAA) *2*

competency *35*

continued operational safety (COS) *95*

continuous analysis and surveillance system (CASS) *86*

continuous monitoring *151–154*

continuous monitoring plan *151*

controlled flight into terrain (CFIT) *24*

corporate knowledge *52, 182*

cost versus benefit *40–45*

crew resource management (CRM) *72–73*

D
Department of Transportation (DOT) *44*

direct costs *42*

E
emergency response plan *16*

evaluation *31*

evidence *162–163*

F
FAA Certificate Management Team (CMT) *27*

Federal Aviation Administration (FAA) *8, 11, 30, 58*
    FAA Airports *7*
    FAA Flight Standards Organization *6*
    FAA Program Office *6*

Five "M" Model *133–138*
    machine *136*
    man *133–135*
    management *136–137*
    media (environment) *135*
    mission *138*

Flight Operations Quality Assurance (FOQA) *61, 64, 154*

Flight Standards District Office (FSDO) *58, 93, 96*

G
gap analysis tool *59–60, 98–100, 113, 190–191*
    detailed gap analysis *99*
    preliminary gap analysis *99*

H

hazard *23*

hazardous attitudes *135*

human error *71*

human factors *4*

Human Factors Analysis and Classification System
   (HFACS) *80*

I

Iceberg of Ignorance *81–82*
   Yoshida, Sidney *81*

indirect costs *42*

internal evaluation program (IEP) *65, 88*

International Civil Aviation Organization (ICAO) *2,
   8–10*
   ICAO Annexes *9–10*
   ICAO Safety Management System Manual *84*
   standards *11*

J

job aid *103, 166–167*
   design validation job aid *106*
   example of *166*
   performance job aids *106*

K

key safety management personnel *50–51, 173*

L

latent failure *72*

Line Operations Safety Audit (LOSA) *64*

N

National Transportation Safety Board (NTSB) *11, 30*

non-punitive employee safety reports *156–157*

Notice for Proposed Rule Making (NPRM) *58*

P

Part 141 pilot school *123–124*

practical drift *84–85*

predictive thinking *54*

processes *50, 80*

process managers *122*

process owners *122*

R

residual risk *154*

responsibility *15*

risk acceptance *25–26*

risk control *2, 139–144*

risk matrix *23–25, 132*
   likelihood and severity criteria *25*

S

safety *70*

safety assurance (SA) *2, 3, 14, 29–35, 150–157, 160–167*
   analysis *32–33*
   data acquisition & processes *30–32*
   preventive/corrective action *33–34*
   system assessment *33*
   system operation/monitoring *29–30*

safety culture *2, 35, 80, 170–178*
   accountability *175*
   commitment to risk reduction *173*
   flexibility *174*
   informing *176–177*
   justness *172*
   learning *174–175*
   management involvement *176*
   openness *172–173*
   training *175*
   use of information *173*
   vigilance *174*

safety management system (SMS) *2*
   creating your SMS manual *182–192*
   defined *50–52*
   four components of *14, 37–38*
   implementation *108–109, 112–115, 190–191*
   management of *85–89*
   safety assurance component *87*
   scalability *62–68*
   versus safety program *50–55, 88*

safety performance *123–126*

safety policy *3, 14–18, 118–126*
   objectives *123–124*
   safety policy statement *118–120, 126, 188–190*
   scalability *62*

safety promotion *2, 3, 14, 35–37*

safety risk assessment (SRA) *21, 132–147*

safety risk management (SRM) *2, 3, 14, 18–29, 130–148,
   151*
   decision making *2*
   hazard identification *22, 131*
   risk analysis *23–24, 131*
   risk assessment *24–26, 131–132*
   risk control *27–28, 132*
   scalability *62–64*
   system description and analysis *19–21, 131*

safety thinking *71*

Senior Technical Specialist (STS) *94*

Service Difficulty Reporting System (SDRS) *156*
SHEL model *82–84*
    liveware *82–84*
SMS Pilot Project *6*
SMS Program Office (SMSPO) *8, 58, 93–94, 96, 104, 112, 125*
SMS Regional Office Point of Contact (RPOC) *93, 96*
SMS Voluntary Program (SMSVP) *7–8, 58, 93–106*
    CMT validation phase *100–102*
    COS phase *105–106*
    documentation validation phase *102–103*
    implementation *95*
    performance demonstration phase *103–104*
    preparation phase *95–100*
    SMSVP Standard *15, 92, 160, 184*
    SMSVP Status Roster *95, 102*
    three levels *8*
SMSVP Active Applicant *8, 95*
SMSVP Active Conformance *8, 95*
SMSVP Active Participant *8, 95, 102*
SMSVP Guide *51, 92, 106, 165–167, 170, 186*
SMSVP non-active participant *95*
Swiss cheese model *71–80*
    inadequate supervision *76–78*
    organizational influences *78–80*
    Reason, James *72*
    unsafe acts *73–75*
system *51*

T
trans-cockpit authority gradient *77*

V
validation project plan (VPP) *100–102*
value of a statistical life (VSL) *43*
    value of preventing injuries *44*
Voluntary Disclosure Reporting Program (VDRP) *30*
Voluntary Safety Reporting Programs (VSRP) *155*